MAGNETISM AND MAGNETIC MATERIALS

Edited by **Neeraj Panwar**

Magnetism and Magnetic Materials
http://dx.doi.org/10.5772/intechopen.68579
Edited by Neeraj Panwar

Contributors

Karine Chesnel, Jurgen Kosel, Hanan Mohammed, Julian Moreno, Rohit Shahi, Rajesh Kumar Mishra, Sachin Shaw, Neeraj Panwar

First published in London, United Kingdom, 2018 by IntechOpen
IntechOpen is the global imprint of INTECHOPEN LIMITED, registered in England and Wales, registration number: 11086078, The Shard, 25th floor, 32 London Bridge Street
London, SE19SG – United Kingdom
Printed in Croatia

British Library Cataloguing-in-Publication Data
A catalogue record for this book is available from the British Library

Additional hard copies can be obtained from orders@intechopen.com

Magnetism and Magnetic Materials, Edited by Neeraj Panwar
p. cm.
Print ISBN 978-1-78923-678-1
Online ISBN 978-1-78923-679-8

For EU product safety concerns:
IN TECH d.o.o., Prolaz Marije Krucifikse Kozulić 3, 51000 Rijeka, Croatia,
info@intechopen.com or visit our website at intechopen.com.

Meet the editor

Dr. Neeraj Panwar, born in town of Ailum of Uttar Pradesh province of India, obtained his Doctorate degree in Physics from Indian Institute of Technology Delhi India. Subsequently, he obtained several years of Post Doctoral research experience from the University of Puerto Rico, U.S.A. and the University of Aveiro, Portugal. Dr. Panwar joined the current job as an Assistant Professor of Physics in 2013 at the Central University of Rajasthan India. He is supervising three Ph.D. students for their thesis and has guided many Masters students for their dissertation. He has several ongoing research projects funded by different agencies in India. He has published sixty research papers in internationally peer reviewed journals. His research area includes magnetism, multiferroics and piezoelectrics.

Contents

Preface

Let Noble thoughts come to us from all the sides

The research in the field of magnetism especially on the magnetic materials has been moving forward by leaps and bounds over the last several decades. This includes research on nanowires thin films, alloys, and ceramics. A need is felt for a single place where one can find the latest update about the ongoing research in magnetism. Through this book, some relevant research work pertaining to magnetism and magnetic materials has been addressed. The first chapter of the book is a glimpse of history of magnetism. The next two chapters elaborate the work on magnetic nanowires and nanoscale magnetic domain memory in ferromagnetic thin films. Chapter four delineates magnetic and magneto-transport properties of Heusler alloys. Chapter five is about the research on magnetic characterization of alloys with high entropy. Finally, last chapter describes mathematical modelling of magnetic drug targeting in microvessel.

I am very much grateful to all the contributing authors of the book. I would also like to thank Mr. Slobodan Momcilovic A&I Operations and Bibliographic Coordinator, IntechOpen publication for providing the technical support whenever needed. I am also indebted to my family, friends and my wife Dr. Indrani Coondoo for providing me every support during this journey of book editing.

Last but not the least, I hope that readers will find the topics covered in this book useful and interesting.

Dr. Neeraj Panwar
Department of Physics
School of Life Sciences
Central University of Rajasthan
Bandarsindri-305817
Kishangarh, Ajmer
Rajasthan, India

Section 1

Introduction

Chapter 1

Introductory Chapter

Neeraj Panwar

Additional information is available at the end of the chapter

http://dx.doi.org/10.5772/intechopen.76447

1. Introduction

1.1. Magnetism and magnetic materials

Magnetism is one of the oldest phenomena known to mankind. For example, iron used to be magnetized by stroking it with another magnet or simply by placing it in the proximity of a strong magnet. The oldest magnetic compound, so-called mineral of magnetite (Fe_3O_4) was initially found in the district of Magnesia of modern Turkey. The word "magnet" is a Greek word. The magnetic properties of materials are entirely governed by the motion of electrons of the atoms. The simplest form of an electromagnet can be produced by wrapping copper wire into the form of a coil and connecting the wire to a battery. A magnetic field is created in the coil, but it remains there only while electricity flows through the wire. There are different types of magnetism existing today. They can be categorized into paramagnetism, ferromagnetism, ferrimagnetism, and antiferromagnetism. The applications of magnetism are many including data storage, magnetic switches, and in the medical field like magnetic resonance imaging (MRI). In 1905, Langevin explained the theory of diamagnetism and paramagnetism but was unable to explain the ferromagnetism. In 1906, the ferromagnetism theory was developed by Weiss. The first commercial steel magnets were made available in 1919 and were quench-hardened steel magnets. In 1930, I. Mishima produced the first Alnico magnet that contained an alloy of iron, aluminum, and nickel. Furthermore, Hermann Kemper studied the use of magnetic fields in conjunction with trains and airplanes. In 1952, J.J. Went et al. invented the first ceramic magnets based on barium, lead-iron oxides, and strontium at the Philips Company. In 1966, rare-earth magnets with high-energy product were reported by Karl J. Stmat. NdFeB magnets have significantly boosted the development of computer peripherals such as voice coil motors and actuators, in both downsizing and enhancing their performance characteristics. For example, in 1984, computer disk drives of ≈10 MB size were in use. Today, external hard drives of much smaller physical dimensions, but with a storage capacity of a few hundred GB or TB, have become common. More importantly, these hard drives are available at affordable costs.

Some of the magnetic materials exhibit the so-called hysteresis behavior between magnetization and the applied magnetic field. Based on the hysteresis curve, the magnetic materials can be further divided into soft and hard magnetic materials. Magnetic materials, like ferromagnets, demonstrate the domain patterns. The domains are separated from each other by a domain wall.

1.2. Modern magnetism

In modern times, magnetism has put forward many peculiar and interesting phenomena in front of the scientific fraternity. The examples include magnetism at nanoscale, spintronics, magnetization reversal, magnetocaloric effect, exchange bias, multiferroicity, and so on. With the advance in the field of material growth techniques, it is now possible to control magnetism up to few nanometer length scale and grow artificially engineered magnetic materials. For example, with the thin films of ferromagnetic materials exhibiting perpendicular magnetic anisotropy and nano-sized magnetic domains, it is possible to produce ultra-high storage density.

Magnetic domain memory (MDM) is the ability for a ferromagnetic film to retrieve its particular magnetic domain pattern after an external magnetic field has been applied and erased the pattern. Furthermore, the single-domain behavior of magnetic nanowires with diameters below 100 nm is experimentally observed with Lorentz force microscopy, magnetoresistance measurements, and magnetic force microcopy. These methods confirm magnetization switching by domain wall motion without any intermediate magnetization states.

Author details

Neeraj Panwar

Address all correspondence to: neeraj.panwar@gmail.com

Department of Physics, Central University of Rajasthan, Ajmer, India

Section 2

Magnetism at Nanoscale

Chapter 2

Advanced Fabrication and Characterization of Magnetic Nanowires

Hanan Mohammed, Julian A. Moreno and Jürgen Kosel

Additional information is available at the end of the chapter

http://dx.doi.org/10.5772/intechopen.71077

Abstract

Magnetic nanowires feature unique properties that have attracted the interest of different research areas from basic physics over biomedicine to data storage. The combination of crystalline and shape anisotropy is mainly responsible for the magnetic properties of the nanowires, whereby different methods for tuning those properties are available. The nanowires typically represent single-domain particles, and magnetization switching occurs via domain walls nucleated at the ends of the nanowire and traversing it. Combined with a high biocompatibility, iron or iron oxide nanowires can be used as nanorobots for biomedical applications, destroying cancer cells, or delivering drugs. The nanowires are also attractive for data storage, especially in a three-dimensional device, because of the high-domain wall speed that has been theoretically predicted. This chapter offers an introduction to the electrochemical synthesis of cylindrical nanowires in anodic aluminum oxide (AAO) templates. Template modification techniques such as barrier layer thinning, barrier layer etching, and diameter modulation are discussed. Advanced fabrication techniques of nanowires with varying structural and chemical variations such as multisegmented and core-shell nanowires are elaborated. The characterization of single nanowires encompassing physical, magnetic, and electrical techniques is covered.

Keywords: nanotechnology, nanofabrication, magnetic nanowires, cylindrical nanowires, anodic aluminum oxide templates, anodization, diameter modulated nanowires, core-shell nanowires, multisegmented nanowires, single nanowire characterization, single nanowire electrical contacting

1. Introduction

Cylindrical magnetic nanowires are novel materials that offer unique properties, mainly due to a high aspect ratio and shape anisotropy. They are typically characterized by a single magnetic domain, rendering them permanent magnetic. This feature allows utilizing such nanowires as remotely operated nanorobots, i.e., induce motion, produce heat, or sense their location. This makes them attractive for numerous applications like flow sensors [1], magnetic separation [2], bio-inspired tactile sensors [3], energy harvesting [4], cancer treatment [2, 5], drug delivery [6, 7], and MRI contrast agent [2, 8]. Nanowires are also used for other applications such as magnetic force microscopy (MFM) tips [9], giant magnetoresistance (GMR) [10], spin transfer torque (STT) [11, 12], and data storage devices [13–15].

In order to exploit these nanowires to their full advantage, a reproducible fabrication method is required that yields nanowires of high quality and specific properties. Electrodeposition of nanowires into templates is such a method and is described in more detail. The properties of magnetic nanowires can be tailored in different ways to optimize them for particular applications. For instance, multisegmented nanowires and core-shell nanowires will be discussed below, which allow combining the properties of several materials. This can result in e.g., large magnetization values in combination with high biocompatibility, as will be shown below.

On the other hand, characterizing these properties is a crucial task for the magnetic behavior of the nanowires and the influence of different parameters. The magnetic properties of nanowires have typically been measured for large numbers of nanowires arranged in arrays inside the templates, which provide large signals but compromise the information content. More recently, individual nanowires were characterized with magnetic force, magneto-optical Kerr effect, and electron microscopy methods as well as magneto resistivity. In this chapter, we will focus on single nanowire characterization methods and results, which provide much deeper insight into the magnetic and structural details of the nanowires. Further understanding can be obtained, when the experimental results are combined with micromagnetic simulations, which conveniently allow varying parameters.

2. Anodic aluminum oxide as nanowire templates

Electrochemical deposition into nanostructured templates represents a widespread and inexpensive bottom up approach for the fabrication of metallic nanowires with a high aspect ratio and high pore density [16–18]. Anodic aluminum oxide (AAO) templates have emerged as a popular template, due to their versatile fabrication technique, wherein the pore diameter, length, and inter pore distances can be tuned with ease. The AAO templates are relatively easy to make in-house with low costs and scalable processing. In addition to obtaining vertically aligned nanopores, modifying the template fabrication process can yield diameter-modulated nanopores [19, 20].

2.1. Mechanism of anodic aluminum oxide template formation

Self-ordered pores organized in a close-packed hexagonal lattice are obtained by anodic oxidation of aluminum (mild anodization) in a suitable acidic environment such as sulfuric, oxalic, or phosphoric and under appropriate electrolyte conditions such as pH, concentration, and temperature [21–25]. The AAO template formation or anodization proceeds with the conversion of the naturally occurring preexisting oxide film on the aluminum surface into a thicker oxide layer referred to as a planar barrier-type film (**Figure 1(1)**). The planar barrier-type film converts into a porous oxide layer by forming random cracks in the outer regions of the barrier oxide layer (**Figure 1(2)**), which further forms individual pathways until a steady-state pore structure is created (**Figure 1(3)**). Once a steady state is achieved, pores proceed throughout the anodized material (**Figure 1(4)**).

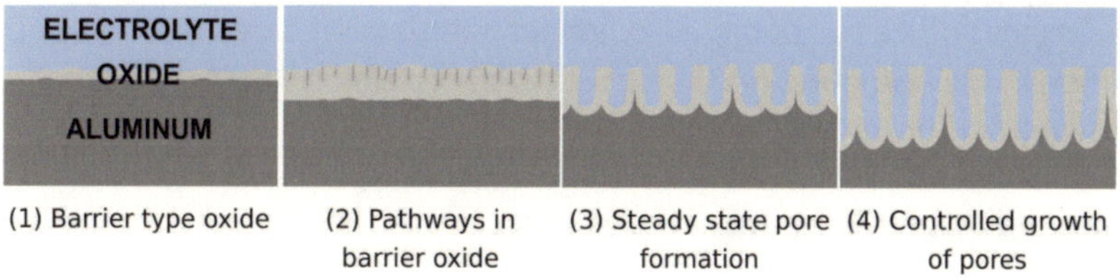

Figure 1. Mechanism of anodic aluminum oxide template formation. (1) Formation of barrier type oxide layer, (2) pathways created in the barrier oxide layer, (3) steady state formation of pores, (4) controlled growth of nanopores.

Since pore formation is initiated at random locations on the aluminum surface, an ordered pore arrangement can be obtained by the fabrication of dimples in the aluminum at the intended position of the pores. The dimples can be fabricated by direct indentation such as with the tip of a scanning probe microscope [26–28], milling with a focused ion beam [28, 29], or nanoindentation using Si_3N_4 [30] or Ni [31] stamps. An easy alternative approach for pore ordering is the two-step anodization or double anodization method developed by Masuda and Fukuda based on self-organization [21]. It is this double anodization method that has attracted much interest for AAO to be used as nanowire templates as well as for many other applications. The porous oxide film formed after an anodization process (first anodization) is poorly ordered with varying pore diameters and inter pore distances at the top surface but has a high degree of order at the bottom of the pores (**Figure 2(1)**). In the double anodization method, the first porous oxide layer is chemically removed (**Figure 2(2)**) to reveal semi-spherical etch pits or dimples (**Figure 2(3)**) that are highly ordered and uniform. A second anodization process is

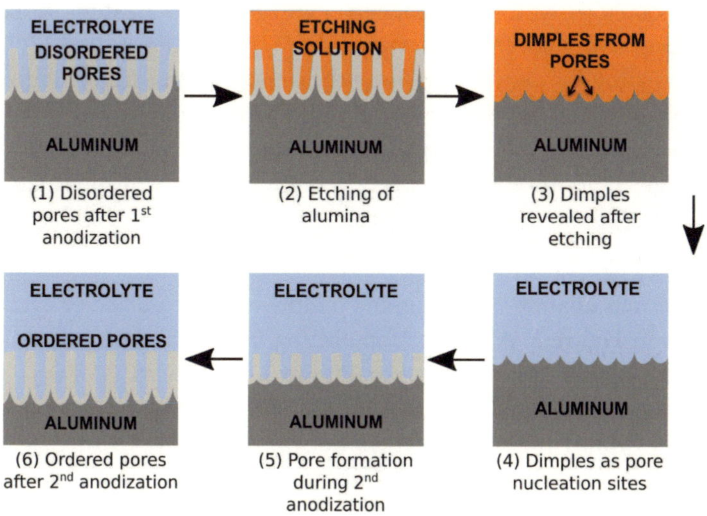

Figure 2. Mechanism of double anodization process. (1) Disordered pores formed after the first anodization step, (2) alumina layer is etched, (3) etching reveals dimples on the aluminum template, (4) the dimples serve as pore nucleation sites for a second anodization process, (5) formation of ordered pores during second anodization, (6) ordered pores obtained after second anodization.

performed, wherein the dimples serve as the pore nucleation sites (**Figure 2(4)**), and the subsequent pores grow exactly on the dimples (**Figure 2(5)**). Thus, the first anodization itself is used as the prepatterning step. Since anodization is a faradic process, the longer the anodization process, the longer will be the pores. Also, longer anodization times result in improved pore ordering, wherein the best pore ordering exists at the bottom of the AAO pores, i.e., the layer that was grown last (**Figure 2(6)**).

2.2. Fabrication of diameter-modulated templates

A diameter-modulated cylindrical nanowire refers to a nanowire, which possesses a changing diameter along its length. Such nanowires can be obtained by using diameter-modulated alumina templates, which are prepared through a combination of *hard* and *mild* anodization processes [19, 32, 33]. Mild anodization of aluminum was described in Section 2.1 and is constrained to very tight equilibrium regimes (voltage, temperature, concentration, and stirring) at which highly ordered, well-defined pores with fixed diameter and inter pore distance are obtained. Deviations from those regimes lead to lack of ordered pores or high current density flowing through the oxide layer, due to high voltages (burning) of the sample. However, the self-ordering pore regimes can be obtained at high voltages with the same concentration of e.g., oxalic acid, provided low temperatures and a starting thick oxide layer. This high voltage anodization is referred to as hard anodization, and in both cases (hard and mild), the anodization current represents a measure of the effective electric field at the oxide layer, and its value is related to the movement of ionic species through the electrolyte/oxide/metal interfaces, the oxygen ions from the electrolyte being the limiting factor [34, 35]. Because of the high-applied voltage in hard anodization, the electric field at the oxide/metal interface results in a continuously increasing thickness of the oxide layer, limiting the length of ordered pores by building up internal stresses, due to the volume expansion. In order to produce diameter modulations, two main approaches can be used: a combination of mild anodization in phosphoric acid (high diameter) and hard anodization in oxalic acid (low diameter), changing solutions in each step, as in [34] or using pulsed [36] or cyclic [37] anodization on the same solution. When using the same solution, hard anodization voltages would yield a higher diameter pore than mild anodization ones. This is due to the faster dissolution of the oxide layer driven by the higher electric field, compared to lower (mild) anodization voltages [35]. Another approach to fabricate diameter-modulated templates is by chemical widening of the pores after a segment of plated metal has been deposited in a nonmodulated anodized template [38].

A general diameter modulation of an alumina template, using pulsed or cyclic anodization, would proceed as in **Figure 3**. After electropolishing the aluminum disc, mild anodization is performed to form a first thick oxide layer (**Figure 3(a)**). This reduces the effective electric field at the oxide layer when higher voltages are applied, limiting the production of high current densities from *burning* the sample. Then, the voltage is slowly increased (0.05–0.5 V/s) until it reaches the hard anodization value (**Figure 3(b)**). At this point, cycles between hard (high diameter) and mild (low diameter) anodization voltages can be repeated to achieve the desired pore length (**Figure 3(c)**).

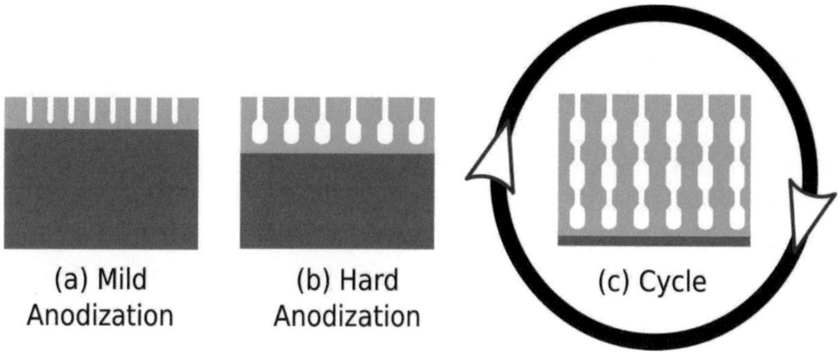

Figure 3. Diameter modulation of an alumina template. (a) Mild anodization process is carried out to create a thick oxide layer to withstand the high voltages of hard anodization, (b) voltage is slowly increased until the hard anodization voltage is reached, (c) changing the voltage from hard to mild anodization values results in the diameter modulation.

3. Electrochemical synthesis of nanowires

Electrochemical deposition or electrodeposition is an inexpensive and widely exploited technique for the deposition of metals through the chemical reduction of metal ions from an aqueous electrolyte [39]. Electrodeposition into AAO templates yields cylindrical nanowires that grow in a bottom-up fashion. The electrodeposition setup consists of an AAO template placed in contact with a cathode, and an anode placed parallel to it in an aqueous electrolyte consisting of the precursor material. Upon the application of an electric field, cations of conductive material diffuse toward the cathode and are reduced therein, resulting in the growth of nanowires inside the AAO pores whose length is monitored by the charge density during deposition. There are two common techniques to deposit nanowires namely direct current (DC) deposition and pulsed deposition. In order to perform electrodeposition, a conductive pathway should exist between the electrolyte and the cathode, but the AAO template obtained after the second anodization process lacks this feature. To overcome this problem, two approaches exist namely barrier layer removal by voltage control [18, 40] and barrier layer removal by chemical etching. The former approach results in aluminum-supported AAO templates, whereas the latter results in free-standing AAO templates.

3.1. Fabrication of nanowires in aluminum-supported aluminum oxide template templates

In the case of barrier thinning approach, the barrier layer formed after the second anodization (**Figure 4(1)**) is reduced by applying a decreasing voltage in successive steps. This process results in the formation of small root-like pathways or dendrites at the bottom of the hemispherical pores (as seen in **Figure 4(2)**), along with the thinning of the alumina barrier layer [41]. The thinned barrier layer along with the dendrites creates a lower potential for the deposition current to tunnel through the alumina barrier layer during the subsequent electrodeposition step (**Figure 4(3)**). The deposition into aluminum supported AAO templates is a straightforward process, as it does not involve any chemical etching post-anodization, unlike the free-standing templates (Section 3.2). The pulsed deposition technique is utilized in these templates, in order to avoid charging of the barrier layer and ensure a uniform deposition of the nanowires. The pulsed deposition consists of applying a negative current pulse which attracts the positive metal ions to the bottom of the AAO pores, followed by a positive voltage pulse which discharges the alumina barrier layer. Finally, a recovery time, wherein neither a current nor a voltage pulse is applied, aids in refreshing the solution at the bottom of the pores.

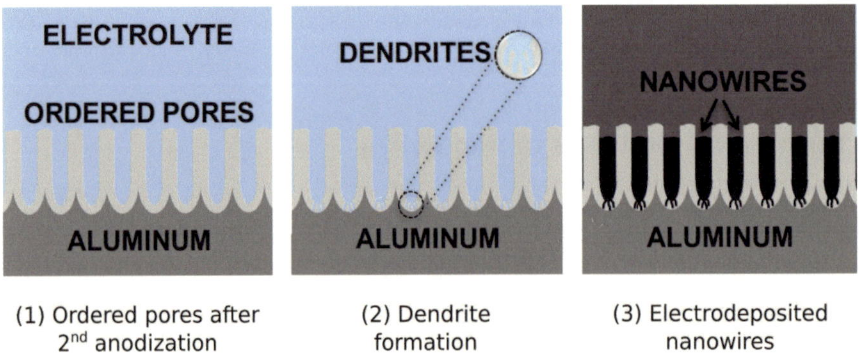

Figure 4. Electrodeposition into an aluminum supported AAO template. (1) Ordered pores after second anodization, (2) barrier layer thinning by voltage reduction resulting in the formation of dendrites, (3) electrodeposition of nanowires.

3.2. Fabrication of nanowires in free standing aluminum oxide templates

In order to obtain a free-standing template, i.e., consisting of pores open from both sides, after the second anodization (**Figure 5(1)**), the nonoxidized aluminum is removed by chemical etching under constant stirring (**Figure 5(2)**). Once the aluminum layer is removed, the alumina layer consisting of nanopores is chemically etched using an aqueous solution (**Figure 5(3)**), thereby resulting in open-ended nanochannels (**Figure 5(4)**). A conducting pathway for the electrodeposition step is achieved by physical vapor deposition of a suitable metal such as gold on the backside of the template to act as an electrode (**Figure 5(5)**). The template is then immersed in the suitable electrolyte (**Figure 5(6)**), and electrodeposition is then performed (**Figure 5(7)**) Once the electrodeposition is completed, the template is rinsed and dried (**Figure 5(8)**), and finally, the electrode is removed by plasma etching. Even though the free-standing templates are not straight forward to fabricate in comparison with the aluminum-supported AAO templates, their pores can be chemically widened, thereby providing more diameter flexibility to the AAO templates postfabrication. In addition to this, DC deposition can be performed due to the good electrical contact between the AAO template and the cathode.

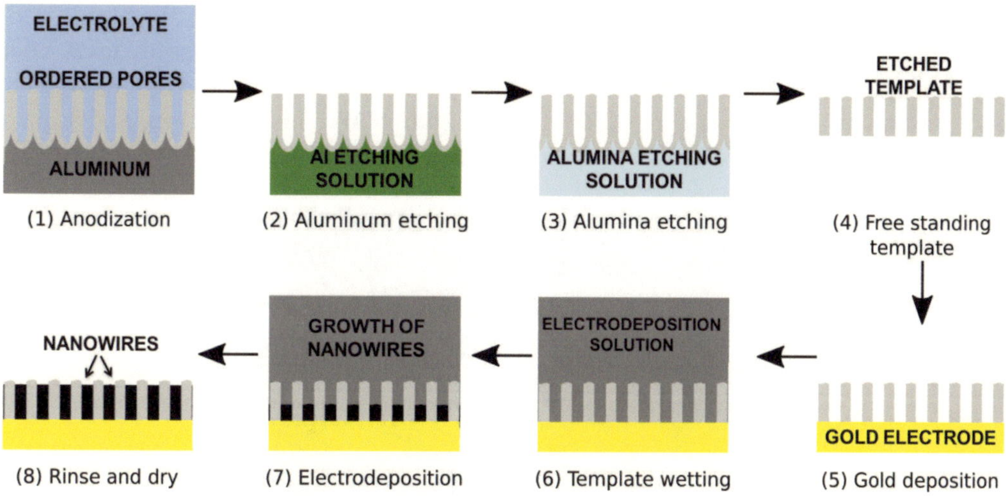

Figure 5. Electrodeposition into free-standing AAO template. (1) Ordered nanopores obtained after second anodization, (2) chemical etching of aluminum layer, (3) chemical etching of alumina layer, (4) free standing AAO template obtained, (5) electrode deposited onto the AAO template, (6) wetting the template with the electrodeposition solution, (7) electrodeposition of nanowires, (8) electrodeposited nanowires.

3.3. Fabrication of multisegmented nanowires

Multisegmented magnetic nanowires can be used for a wide range of applications such as data storage by using multisegmented Co/Ni nanowires [15, 42], functionalization, and release of molecules [43, 44], cell detection [43], etc. Multisegmented nanowires can be fabricated by electrodeposition into AAO templates using either a single electrolyte solution or more than one electrolyte solutions [15, 45–47]. In the case of a single electrolyte solution, each component within the electrolyte will have differing concentrations and deposition voltages, and alternating between the voltages will result in a multilayered structure. Since both applied voltages have the same sign, what prevents the deposition of both materials at the same time is the difference in concentration. Thus,

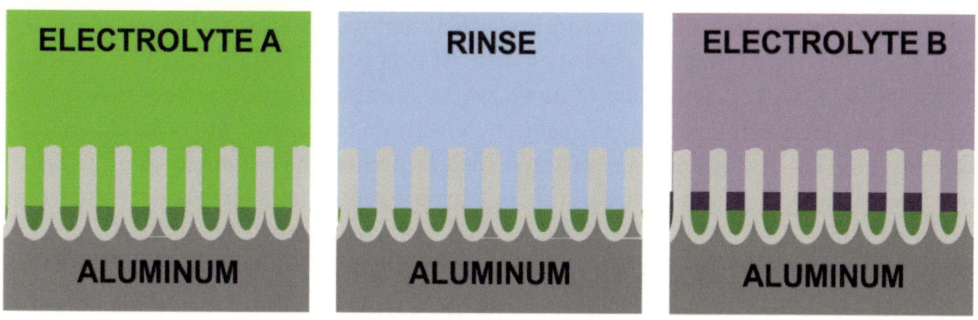

Figure 6. Electrodeposition of multisegmented nanowires using two electrolyte solutions. (1) Deposition of metal using electrolyte A, (2) rinsing of the nanopores, (3) deposition of metal using electrolyte B.

the lower concentration component will be limited by diffusion, when the higher concentration component is being deposited. If different electrolyte solutions are used, after depositing one metal (**Figure 6(1)**), the AAO template is rinsed (**Figure 6(2)**), and the second electrolyte is used for the next deposition (**Figure 6(3)**). This can be repeated several times to obtain a specific number and arrangement of segments [15].

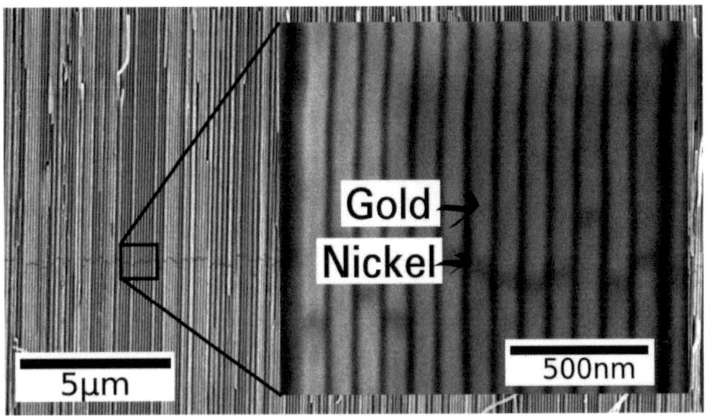

Figure 7. Scanning electron microscopy (SEM) image of a cross-section of an AAO template with multisegmented Au/Ni/Au nanowires. Inset depicts magnified image (reprinted with permission from Moreno [48] copyright 2016 KAUST).

Figure 7 depicts an example of electrodeposition using separate nickel and gold electrolyte solutions with a rinse step in between solution changes. The different contrast exhibited by the nickel segment assists in identifying different composition variation along the nanowire. The variation of the thickness of the nickel segment between different nanowires is a consequence of slight pore diameter variations within the AAO template.

3.4. Fabrication of core-shell nanowires

Core-shell nanowires consist of different core and shell materials such as metal-nonmetal or metal-metal nanowires [49–51]. They are fabricated in order to tune their magnetic properties and take advantage of the combined properties of the core and shell materials [52, 53]. One technique for realizing core-shell nanowires is by utilizing atomic layer deposition (ALD) to fabricate a shell structure. ALD is performed on AAO templates and the thickness of the deposited film is tuned by the number of ALD cycles. Once the shell structure is obtained, the core is fabricated by electrodeposition of the required material. Even though ALD can provide a conformal shell structure, the technique is an expensive as well as challenging process due to the high aspect ratio of the templates. A facile low-cost technique for fabricating metal-oxide core-shell nanowires such as iron-iron oxide (magnetite) nanowires is by oxidizing released iron nanowires in an oven at 150°C, for 10 min to 72 h in an ambient atmosphere (**Figure 8**) [50]. This technique easily allows the tuning of the geometry (such as core radius, shell thickness, and length) and crystallinity (single or polycrystalline core) of the nanowire. Consequently, the magnetic properties can be tuned, i.e., the saturation and remanence magnetization decrease with increasing shell thickness.

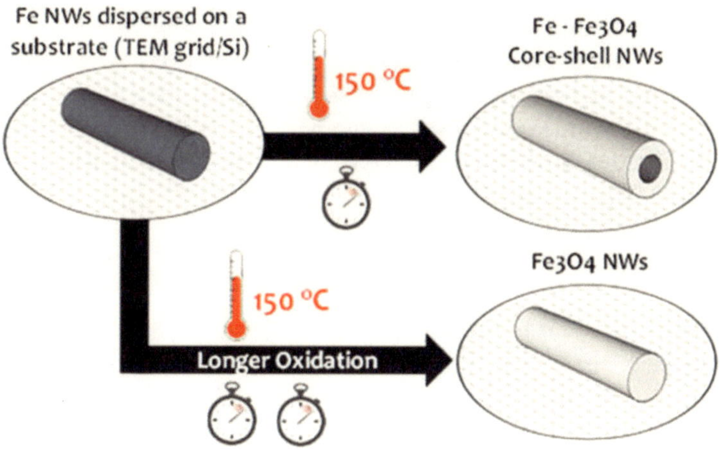

Figure 8. Fabrication process of core-shell Fe-Fe_3O_4 nanowires [18].

4. Characterization of single nanowires

A comprehensive characterization of nanowires is crucial toward utilizing these novel structures for various applications such as biosensors, drug delivery, data storage, etc. Mostly, the collective behavior of nanowires in an array has been studied [54–56], which provides some insight into the magnetic properties of the nanowires but is affected by magnetostatic interactions between them. Hence, single nanowire characterization is desirable, but there are various challenges associated with it, due to the nanowire's small dimensions. For example, the high aspect ratio of the nanowires makes it prone to mechanical deformations during their release from the template as well as during subsequent characterization. The small volume of the nanowire makes magnetic and electrical characterization challenging, due to the low magnetic signal associated with it, as well as its sensitivity to electrical discharges resulting in melting the nanowire. The characterization of single nanowires can be broadly divided into physical, magnetic, and electrical characterization. In this chapter, an overview of the various techniques for single nanowire probing is discussed.

For probing single nanowires, firstly the nanowires have to be released from the AAO template. The nanowires are released by dissolving the alumina using a selective chemical etchant. The process of release varies between nanowires in aluminum-supported AAO templates and free-standing templates with an electrode back layer. The nanowires in the aluminum-supported AAO templates are released by placing the electrodeposited AAO template in a micro centrifuge tube with a solution of sodium hydroxide at room temperature [5]. This dissolves part of the alumina resulting in the detachment of the aluminum back layer, which is then removed from the tube. To further dissolve the alumina, sodium hydroxide at room temperature or a solution of chromium oxide and phosphoric acid at around 40°C is used. Once the alumina is completely etched, the nanowires, which are now suspended in the etchant, are extensively rinsed and stored in high-purity ethanol. The ethanol rinsing is performed by placing the micro centrifuge tube in a magnetic tube holder rack. The nanowires fabricated using the free-standing AAO templates do not possess an aluminum back layer but instead an electrode usually gold is present, which can be easily removed by ion milling. The AAO template is then chemically etched using either sodium hydroxide or a solution of chromium oxide and phosphoric acid, as described previously. The etchant concentrations, temperatures, and times may vary depending on the goal of the characterization studies [57].

4.1. Physical characterization

The geometry, morphology, chemical, and structural characteristics of single nanowires have been widely studied using electron microscopy techniques such as scanning electron microscopy (SEM) and transmission electron microscopy (TEM) [58]. For investigation under an electron microscope, the released nanowires are dispersed onto a substrate such as silicon (**Figure 9(a,b)**) or TEM grid (**Figure 9(c,d)**). The SEM is generally utilized to probe the nanowire dimensions and structural variations along its length. The SEM is particularly useful in magnifying features such as dendrites, which are present at one end of the nanowire, when they are fabricated using the aluminum-supported AAO template (**Figure 3(1)**). **Figure 9(b)** depicts the SEM image of such a nanowire (polycrystalline fcc nickel nanowire) with a relatively smooth topography, which constitutes two parts: the main part, which is cylindrical and an additional dendritic region [41]. A dendrite typically consists of two to four root-like structures of approximately 500 nm length and a slightly varying shape from one nanowire to another. The TEM on the other hand provides a high-resolution image and more detailed morphological information such as surface roughness (**Figure 9(c)**). The TEM allows imaging of the cross-sectional shape of the nanowire, which is ideally expected to be cylindrical, but in some cases, it deviates from the ideal structure (**Figure 9(d)**). This deviation is a consequence of noncircular pore shapes in the AAO template. Both SEM and TEM can provide information on the chemical composition of the nanowire using techniques such as energy-dispersive X-ray spectroscopy (EDS) [59] and electron energy loss spectroscopy (EELS) [60]. Even though the EDS technique can be performed in an SEM, its resolution is limited to identifying the presence or absence of an element, whereas EDS mapping in a TEM provides a higher spatial resolution. In comparison to the EDS, the EELS technique is a challenging technique performed using a TEM but provides a more detailed chemical mapping of the nanowire.

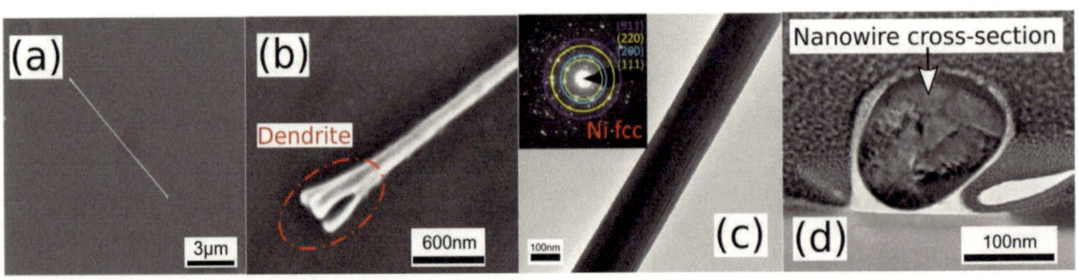

Figure 9. SEM image of a (a) nickel nanowire, (b) magnification of the end of the nanowire depicting the dendritic portion. TEM image of (c) a nickel nanowire with inset of SAED image depicting fcc structure, (d) cross-section of a nickel nanowire (reprinted from Vilanova Vidal et al. [41], with the permission of AIP publishing).

The crystalline structure of a nanowire can be probed using an electron diffraction technique referred to as selected area electron diffraction (SAED) [59]. This powerful technique offers structural information from even a small region of the nanowire. In addition to this, defects such as planar defects can be probed using dark-field TEM.

4.2. Magnetic characterization

Magnetic characterization of single nanowires yields fundamental information such as the saturation magnetization, coercivity, switching field, squareness of individual nanowires, magnetization reversal process, and the types of domain walls involved, as well as the propagation of domain walls within the nanowire. Even though the small nanowire volume makes magnetic characterization challenging due to the low magnetic signal, as well as proper positioning and orientation of the magnetic field with respect to the nanowire axis, various techniques such as magnetometry and microscopy allow the magnetic probing of individual nanowires.

4.2.1. Magnetometry

Magnetometry techniques such as super conducting quantum interference device (SQUID) and magneto-optical Kerr effect microscopy (MOKE) give an insight into the magnetic behavior of a nanowire by the acquisition of a magnetic hysteresis loop. The SQUID magnetometer, which is commonly used for nanowire array characterization, can be modified by lithographically patterning a micro-SQUID detector around a single nanowire in order to sense the low magnetic signal arising from a single nanowire [13, 61]. By measuring the hysteresis curves along the axial and transverse directions, the easy axis of the nanowire can be easily determined. The MOKE microscopy as the name suggests is based on the Magneto-Optical Kerr effect, wherein polarized light reflected from a magnetic material undergoes a rotation of its polarization. Measuring the rotation or change of polarization of the reflected beam gives information about the magnetization state of the sample [62]. This rotation depends on the relative orientation of the incident polarization with the sample magnetization. Therefore, the Kerr effect can be classified into longitudinal, transverse, and polar Kerr effect based on the direction of the magnetization vector with respect to the incident and sample planes. The MOKE microscopy is a versatile and well-established technique and is a more simple approach in comparison to the micro-SQUID, as the samples can easily be measured without an additional structure fabrication. Measuring the Kerr signal from a cylindrical nanowire is however a challenging task due to the small probing area available. In addition to this, the curvature of the nanowire further reduces the signal from the probed area, since at a given time only one Kerr effect (say longitudinal Kerr effect wherein the magnetization is in the plane of the sample) is probed. Due to these reasons, it is crucial to achieve a good focus of the laser spot on the nanowire and attain the highest possible alignment of the setup (such as lenses, laser, etc.), in order to obtain a high Kerr signal from the nanowire.

The MOKE measurements of a Ni nanowire yield a magnetic hysteresis loop as depicted in **Figure 10** [41]. The obtained hysteresis loop is characterized by two saturated states indicating opposite directions of magnetization within the nanowire, with a sudden jump in the hysteresis loop indicating a sharp reversal of the nanowire's magnetization. The magnetic field value at which a sharp reversal of the nanowire magnetization is observed is considered as the switching field (Hsw). The above observations of the magnetization reversal are indicative of a single magnetic domain behavior of the nanowire [41, 63].

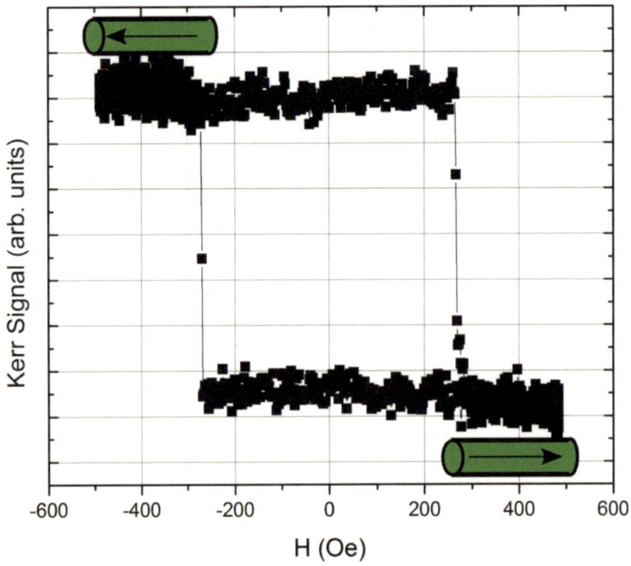

Figure 10. Hysteresis loop of a single Ni nanowire (160 nm diameter and 12 μm in length). The arrows in the schematics indicate the magnetization direction within the nanowire (reprinted from Vilanova Vidal et al. [41], with the permission of AIP publishing).

In addition to obtaining information directly related to the magnetic hysteresis loops of a single nanowire, the magnetization reversal mechanism of the nanowire can be probed by modifying the measurement routine such as by changing the angle of the applied field during the measurement, i.e., angular dependence studies, which would yield the angular nanowire coercivity. Also, the measurement of several individual nanowires and their behavior with fields applied at various angles can be probed in the MOKE [41].

4.2.2. *Magnetic imaging*

Imaging the magnetic signal of a nanowire and its response to an external magnetic field yields useful information about the nanowire's magnetization reversal process, as well as details of the magnetic domains and domain walls. The versatility of the imaging techniques has increased interest in the study and control of domain-wall dynamics, which is fundamental for domain-wall propagation–based applications. Even though imaging of cylindrical nanowires is relatively straight forward, the interpretation of the obtained data is challenging, as the nanowires are three dimensional unlike in the case of planar structures, thus increasing the complexity of single nanowire characterization. Also, the nanowires' cylindrical shape makes the domain-wall internal structure difficult to visualize, like the core of a vortex domain wall.

Magnetic force microscopy (MFM) is a scanning probe microscopy technique which is widely used for the magnetic imaging of nanostructures [64]. The MFM is a special operational mode of the atomic force microscope (AFM), wherein the standard AFM tip is replaced with a magnetic tip usually coated with a high coercivity thin film to fix the magnetization of the tip during imaging. During an MFM scan, the magnetic tip interacts with the magnetic stray fields near the sample surface, and the strength of the local magnetostatic interaction results in a magnetic map of the scanned area. **Figure 11(a)** depicts the AFM image of a Ni nanowire, whose corresponding MFM images at remanence are depicted in **Figure 11(b ,c)**. Firstly, a magnetic field is applied to saturate the nanowire in a particular direction. When the nanowire is magnetically saturated, a dark and bright signal or spot can be observed at the ends of the nanowire. These spots are a result of the stray fields from the ends of the nanowire, which interact with the MFM tip resulting in a repulsive and attractive interaction. In order to understand the magnetization reversal and determine the switching field of the nanowire, the following procedure is performed: firstly, a magnetic field is applied to saturate the nanowire in a particular direction. The magnetic field is then removed, and the nanowire is imaged at remanence. The scans are performed at remanence in order to avoid effects of the stray field of the MFM tip on the nanowire's switching field. The magnetic field is then increased in the opposite direction, until the magnetization of the nanowire is reversed. MFM studies reveal that the nanowire consists of a single magnetic domain with a sharp transition between the two magnetic states. The magnetic field value for which a change in contrast at the ends of the nanowire is observed is considered as the Hsw of the nanowire. The observations from the MFM images agree with the previously obtained MOKE results (**Figure 10**) and confirm the single domain structure of the nanowire [41].

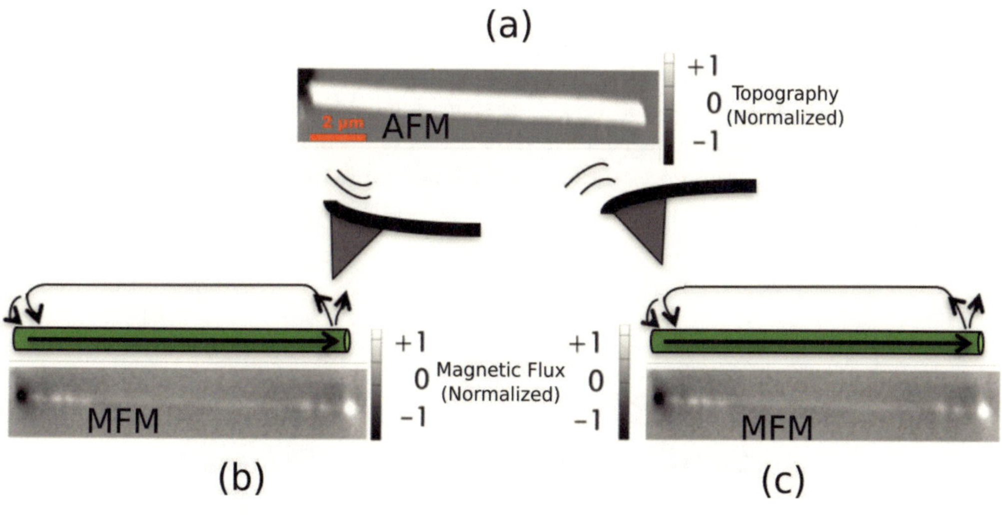

Figure 11. (a) Atomic force microscopy (AFM) image of a Ni nanowire, and (b–c) corresponding magnetic force microscopy (MFM) image of the Ni nanowire. The direction of nanowire magnetization is indicated by the arrows within the nanowire. The interaction of the MFM tip with the nanowire's stray field is depicted as a repulsive and attractive interaction (reprinted from Vilanova Vidal et al. [41], with the permission of AIP publishing).

Several other techniques exist such as Lorentz microscopy [65], modified differential phase contrast microscopy [66], photoemission electron microscopy (PEEM) combined with X-ray magnetic circular dichroism (XMCD) [67, 68], and electron holography [69], which are powerful characterization tools that can yield highly valuable local magnetic information.

4.2.3. Micromagnetic simulations

Even though experimental magnetic characterization techniques offer valuable information about the magnetic nature of the nanowire, they are limited mainly by the 2D data that is obtained as well as the challenges in studying different parameters. Micromagnetic simulation is a versatile tool to understand the magnetization reversal process in nanowires and therefore very useful for interpreting the data obtained from experimental techniques. There are several open source and commercial micromagnetic simulation packages available such as MAGPAR [70], NMag [71], OOMMF [72], LLG Micromagnetics Simulator [73] etc. Studies using micromagnetic simulation tools have revealed that the magnetization reversal in cylindrical nanowires occurs by the nucleation and propagation of domain walls rather than by means of coherent rotation or curling (as was previously understood) and that the type of the domain wall depends on the diameter of the nanowire. As an example, the magnetization reversal process of a polycrystalline Ni (diameter of 160 nm) nanowire was simulated using the MAGPAR package [41]. Since the nanowire consists of a polycrystalline structure, a tetrahedral mesh was randomly assigned with an average mesh size of 6 nm, which is approximately the exchange length of Ni [74]. **Figure 12** gives an overview of the reversal process and consists of four snapshots from the reversal process of the simulated nanowire (green arrows indicate the magnetization direction). Firstly, the nanowire is saturated in the $+M_Z$ direction. The magnetic field is then reduced until remanence after which it is applied in the opposite direction, i.e., $-M_Z$ direction. At remanence, most of the nanowire is magnetized in the $+M_Z$ direction, except for the nanowire ends, which consist of a pair of open vortex areas (OVAs) that are formed as a consequence of the demagnetizing field (to minimize the magnetostatic energy). The magnitude of the magnetic field is increased in the $-M_Z$ direction, and at the switching field, two domain walls (Bloch-point domain walls) nucleate and propagate toward the center of the nanowire, reversing the magnetization direction (Switching process in **Figure 12**). Upon further increase of the magnetic field, the nanowire's magnetization attains a saturated state (saturation state in **Figure 12**).

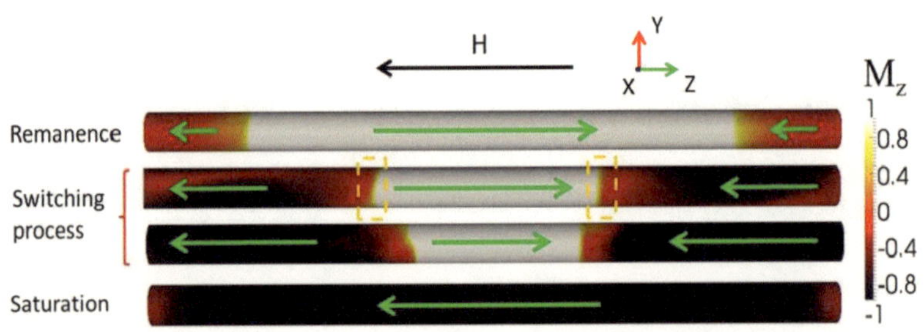

Figure 12. Visualization of the magnetization reversal process from remanence in the +z direction to saturation in the −z direction. The dashed boxes indicate the Bloch-point domain walls (reprinted from Vilanova Vidal et al. [41], with the permission of AIP publishing).

4.2.4. Combination techniques

By utilizing various magnetic characterization techniques (such as magnetometry, magnetic imaging, and micromagnetic simulations) in combination with different measurement routines such as angular dependence studies, it is possible to gain valuable additional information. For example, the effect of the cross-section shape of the nanowire with the switching field (H_{SW}) is investigated using MOKE, MFM, and the Magpar package [41]. The simulations were performed utilizing a circular and a real cross-sectional shape of a nanowire as depicted in **Figure 13(a)** from θ = 0° to 70°. In **Figure 13(b)**, the simulated angular dependencies of the ideal circular and real cross-sections are plotted along with the experimental MOKE and MFM results. It was observed that the values of H_{SW} of nanowires with real and ideal cross-sections

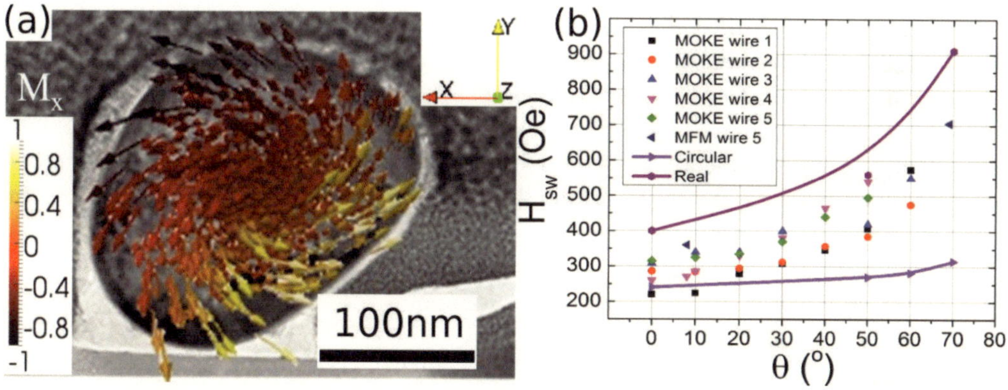

Figure 13. (a) Micromagnetic simulations performed using a real cross-section of a Ni nanowire, (b) dependence of switching field with angle between the nanowire and applied field, of five nickel nanowires measured using MOKE, MFM, and micromagnetic simulations. The simulations were performed using different nanowire cross-section shapes namely circular and real cross-section (reprinted from Vilanova Vidal et al. [41], with the permission of AIP publishing).

are different from each other. For example at θ = 0°, the H_{SW} of ideal cross-section is around 60% of the real cross-section. The angular H_{SW} simulations of the real and circular nanowire cross-section depict two different slopes of the curve. In the real cross-section, increasing angles leads to a higher value of H_{SW}, whereas not much of an increase is observed with increasing angles in the circular cross-section. Interestingly, the H_{SW} values obtained from the circular and real cross-sections define a range inside which all the experimental data fit. This suggests that the different H_{SW} values found for different nanowires are a result of their slightly different cross-sectional shapes. While the simulation with the circular cross-section yields the smallest possible values for H_{SW}, it might be possible to observe even larger values than the ones found for the real cross-section, depending on the actual cross-section shape of the nanowire ends. Thus, the micromagnetic simulations performed here are able to reproduce the experimental scattering obtained for H_{SW} of different nanowires. The cross-sectional shape of the nanowire strongly influences H_{SW}. This is an observation that is particularly relevant from a technological point of view, where the properties of individual nanowires need to be taken into account for device design.

4.3. Electrical characterization

The study of the magnetic and transport properties of cylindrical nanowires such as thermal and resistivity studies, anisotropic magneto resistance, spin transport, etc. is important from a fundamental point of view to investigate unique properties that may arise due to the size of the structures. From a technological perspective, electrical characterization is crucial for device applications. There are several challenges involved in the electrical characterization of a single nanowire such as the contacting of a single nanowire, attaining a low contact resistance, and avoiding electrical discharge that would otherwise result in destroying the nanowire.

4.3.1. In-situ characterization

There are two approaches to electrically address a single nanowire [15, 75–79]. The first approach involves addressing nanowires that are still in the AAO template, i.e., *in-situ* characterization [75, 76]. This is achieved by patterning electrodes on either side of the AAO template, taking care to contact only a single nanowire. The disadvantage of this approach is that the contacted nanowire could be influenced by the neighboring nanowires. Also, it is not possible to analyze the contacted nanowire for the presence of defects or other irregularities which would affect the electrical characterization of the nanowire. In addition to this, the technique is limited to two-point measurement.

4.3.2. Ex-situ characterization

The second approach to electrically address a single nanowire is by contacting released nanowires, i.e., *ex-situ* characterization [15, 76–79]. The advantages of the *ex-situ* approach are that nanowires can be chosen selectively under a microscope, which allows to probe any structural deformities along its length. In this approach, measurements are not limited to two-point geometry, and different sections of the same nanowire (as small as 500 nm) can be probed. An added advantage of the *ex-situ* approach is that several characterization techniques such as MFM and MOKE can be performed on the same nanowire, thus allowing the nanowire's extensive electrical and magnetic characterization [41]. The *ex-situ* characterization can be realized by two methods, by dispersing nanowires onto prefabricated electrodes or by dispersing nanowires onto a substrate (such as Si/SiO_2) and patterning electrodes onto the dispersed nanowire. The former technique requires careful alignment of the nanowires with the electrodes, such as by applying an external magnetic field. The latter does not require any nanowire manipulation but involves the identification of a dispersed nanowire such as by using a scanning electron microscope (SEM) followed by marking its location by patterning alignment marks using a focused ion beam (FIB). Here, it is crucial to obtain sufficiently isolated nanowires which are spread over the substrate. This can be achieved by diluting the nanowire-containing ethanol stock solution to reduce the concentration of dispersed nanowires. The alignment marks aid in electrode placement in the subsequent lithography steps. This approach allows the selection of a nanowire with high precision, as well as contacting only a single nanowire with no additional structures in the vicinity of the nanowire. Generally, the electrode design is patterned either by electron-beam lithography (EBL) [15] or optical lithography [80]. Both the exposure systems have their own advantages and disadvantages. Even though EBL is a labor, cost, and time-intensive process, it

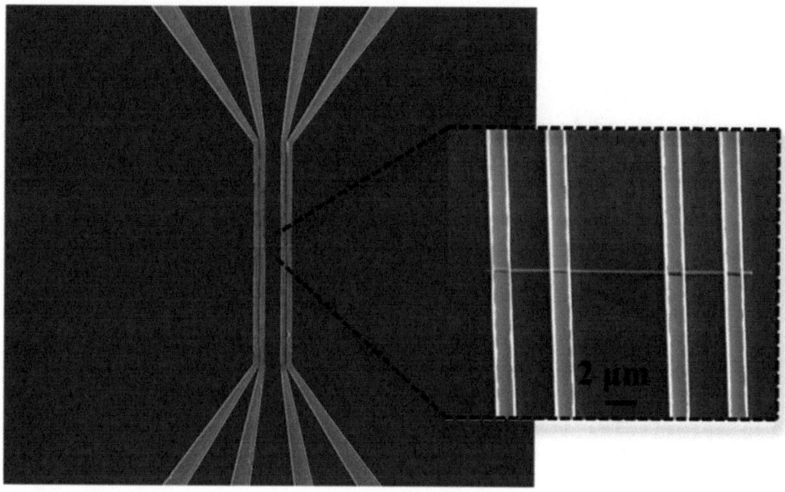

Figure 14. SEM image of electrodes patterned using electron-beam lithography onto a single nanowire. Inset depicts a magnified image (© [2016] IEEE. Reprinted, with permission, from Mohammed et al. [15]).

achieves high precision and small electrode widths (**Figure 14**). This is advantageous when the nanowires are of a shorter length. In the case of EBL, the alignment of the electron beam with the nanowire is extremely challenging especially when it involves a large number of electrodes. Optical lithography on the other hand is a relatively quick process but has limitations regarding the minimum width of electrodes as well as the length of the nanowires that can be used. Once the design is lithographically patterned, electrodes are deposited onto the nanowire. To obtain a good ohmic contact between the electrodes and the nanowire, an etch step is performed prior to the electrode deposition, in order to remove the nanowire oxide layer at the resist openings. This etching process is optimized by systematically controlling the etch time to avoid over or under etching of the nanowire.

4.3.3. Magnetoresistance of individual nanowires

The magnetoresistance (MR) effect has its origin in spin–orbit interaction and depends on the relative orientation of the magnetic moments with respect to the direction of applied electric current. The magnetoresistance (MR) curve offers an insight into the switching behavior of nanowires [81]. **Figure 15(a)** depicts the schematics for MR measurement and (b) shows the magnetization reversal/MR curve of a Ni nanowire (160 nm diameter). At magnetic saturation, the magnetic moments are aligned parallel to the direction of current indicated by the high resistance state in the MR curve. Reversal of the magnetic field results in a decrease of resistance as the magnetic moments continuously rotate away from the direction of current [13, 15, 82]. At a distinct value of the magnetic field referred to as the switching field H_{SW}, there occurs an abrupt reversal of the magnetic moments, which is indicated by a sudden jump of resistance [15], as shown in **Figure 15(b)**.

The MR curve is also useful to identify interruptions in the domain wall motion during the magnetization reversal of the nanowire. These interruptions referred to as pinning can

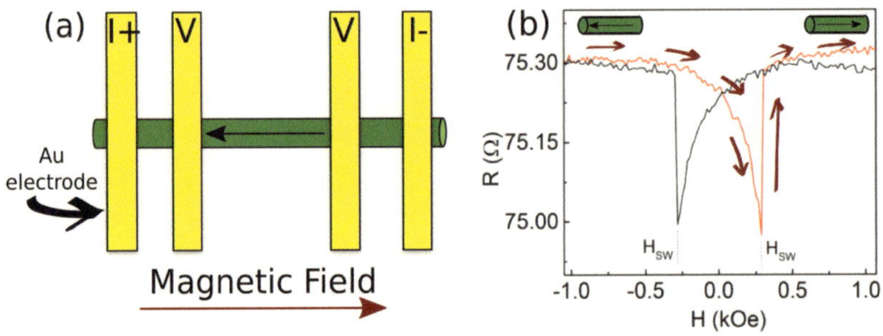

Figure 15. (a) Magnetoresistance measurement schematic depicting a nanowire with four electrodes patterned onto it. The black arrow within the nanowire indicates the nanowire's magnetization direction, (b) Magnetoresistance curve of a Ni nanowire starting from negative saturation to positive saturation (depicted by the red arrows) followed by performing the measurement in the reverse direction, i.e., from positive to negative saturation. The black arrow within the nanowire indicates the nanowire's magnetization direction (reprinted from Vilanova Vidal et al. [41], with the permission of AIP publishing).

be a consequence of a defect (pinning site) within the nanowire. Pinning can be artificially introduced by local modification of magnetic properties by utilizing multisegmented or diameter-modulated nanowires, as well as by introducing notches onto the nanowire. For nanowire-based data storage applications, it is essential to create reliable pinning sites along the nanowire so that domain walls (or bits) can be reliably pinned and depinned (to move the bits) along the nanowire.

Figure 16(a) is an SEM image of a nickel nanowire (20 μm long with a diameter of 160 nm) with four triangular notches patterned onto it using an SEM equipped with a focused ion beam. The MR curve of the notched nanowire constitutes a nonabrupt reversal or a step in the magnetization reversal curve at the H_{SW} (~150 Oe). This discontinuity during the magnetization reversal is an indication of a domain wall which would only depin upon further increase of the applied field. The magnetization reversal at H_{SW} shows two steps (the circles represent the step positions) in **Figure 16(b)**. These two steps are present in both directions and are an indication for domain wall pinning at the two constriction sites.

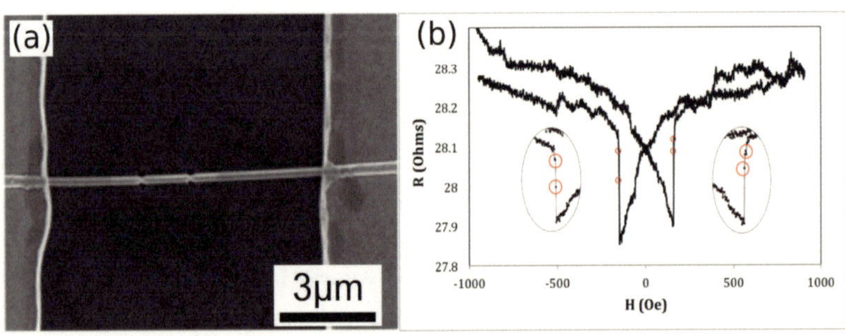

Figure 16. (a) SEM image of a Ni nanowire with notches (to pin domain walls) etched using a focused ion beam, (b) MR curve of the Ni nanowire with notches. Insets depict the magnified area of the MR curve where pinning is observed (represented by the circles).

5. Characterization of multisegmented Co/Ni nanowire

The characterization of multisegmented Co/Ni nanowires (consisting of hcp Co and fcc nickel structure) reveals interesting features. It has been observed that the interface between the Co and the Ni segment act as pinning sites (**Figure 17**). This is an important feature for the application of such nanowires as 3D memory devices.

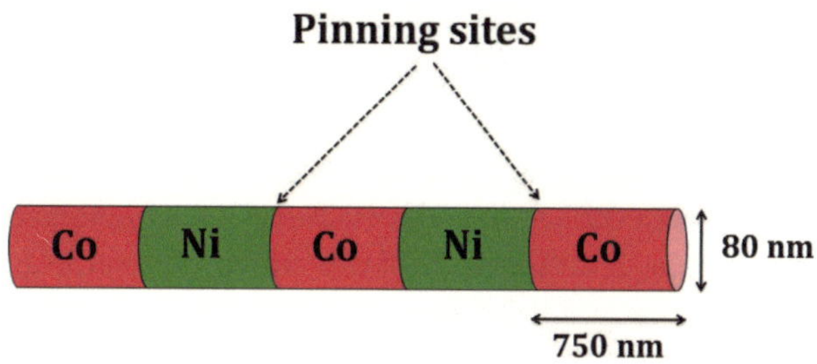

Figure 17. Schematic of a multisegmented nanowire consisting of alternating segments of cobalt and nickel.

Figure 18 depicts the MR curve of a multisegmented Co/Ni nanowire. The magnetization reversal of the nanowire proceeds from a magnetically saturated state indicated by the high resistance value in the MR curve (**Figure 18(a)**). As the direction of applied field is reversed, a gradual decrease in resistance is observed. This decreasing resistance in the MR curve is a consequence of the rotation of the nanowire's magnetic moments away from the direction of applied current. We

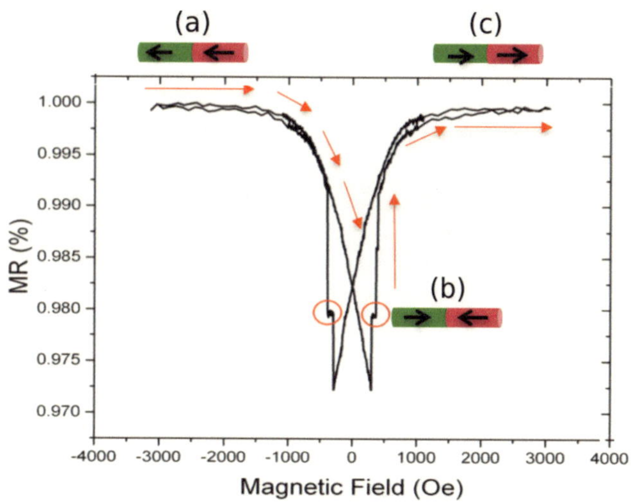

Figure 18. Magnetoresistance curve of a multisegmented co/Ni nanowire. Stage (a) of magnetization reversal depicts the saturated state wherein the magnetic moments are aligned parallel to the current, stage (b) depicts a nanowire with a domain wall (head to head in this case), and finally, stage (c) represents the saturated state with the nanowire having a reversed magnetization direction to that in stage (a). (reprinted (adapted) with permission from Ivanov et al. [83]. Copyright (2017) American Chemical Society).

find that the MR curve of the multisegmented Co/Ni nanowire differs from that of the Ni nanowire in **Figure 15(b)**. Here, the MR curve is similar to the nanowire with notches (**Figure 16(b)**), i.e., instead of the distinct jump at H_{SW}, it displays an interruption or plateau during the jump (**Figure 18(b)**). The plateau indicates that the magnetization does not reverse at once, and magnetization reversal proceeds only upon further increase of the applied magnetic field. This region of the curve is indicative of the pinning of a domain wall, which continues to propagate upon further increase in the magnetic field. Finally, the direction of magnetization within the nanowire is reversed, leading to a parallel alignment of the magnetic moments with respect to the current, which is indicated by the high resistance value in the MR curve (**Figure 18(c)**).

MFM studies performed on the multisegmented Co/Ni nanowire reveal interesting features. As discussed earlier in Section 4.2.2, the MFM image of a single domain state of a nanowire is characterized by a bright and a dark spot at the ends of the nanowire (**Figure 11**). The MFM image of a Co/Ni nanowire with a domain wall pinned at the interface however displays a different stray field configuration. In this case, the contrasts or spots from the ends of the nanowire would be the same, and in addition to this, an alternate contrast would be observed at the interface. **Figure 19** is an MFM image of a multisegmented Co/Ni nanowire and is characterized by two bright spots at the end of the nanowire and a more pronounced black spot at the interface, from which the magnetization state of the nanowire a two domain state with a domain wall pinned at the center can be inferred. The MFM image confirms that the interface in fact acts as a pinning site as indicated by the MR curve.

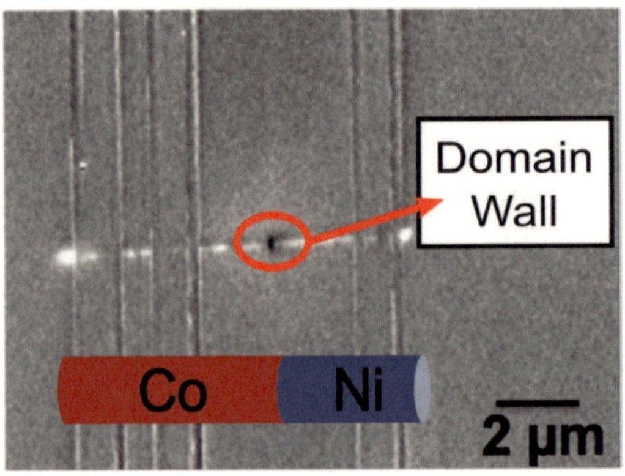

Figure 19. MFM of a multisegmented co/Ni nanowire with a domain wall pinned at the interface. The arrows within the nanowire schematic represents the direction of magnetization within the nanowire (© [2016] IEEE. Reprinted, with permission, from Mohammed et al. [15]).

6. Characterization of diameter modulated nanowires

Nanowires grown in templates allow for the fabrication of nearly one-dimensional structures (see Section 3.2 and 3.3). When the shape of the pores is tailored, e.g., by diameter modulation as described in Section 2.2, the growing nanowires will fill the pores and acquire the geometry of the AAO template (**Figure 20(a)**). From the various experimental and micromagnetic simulation results, general behaviors of the magnetization reversal mechanisms can be found that relate to the nanowire diameter modulation. The nucleation of a domain wall begins at the free ends and/or at the diameter transition, and the domain wall propagates from the thicker to the thinner segment [84]. Small diameters favor transverse domain walls, due to the exchange interaction contribution, while larger diameters favor vortex domain walls, where longer range dipolar interaction becomes more relevant [85]. This means H_{SW} values are dominated by the segment of larger diameter and explain why small differences in diameter between the two ends of a nanowire promote the propagation of a domain wall from only one end. Also, the dipolar interaction between segments with different diameters reduces the switching and saturation fields compared to a conventional nonmodulated nanowire with diameter equal to the thinner segment [16, 32, 84]. This is due to the strong interaction between the stray fields at the ends of each segment which is a result of the sudden diameter change. Domain walls can nucleate at these perpendicular surfaces as a mechanism to minimize the magnetostatic energy [86].

Figure 20(b) shows an AFM image of a FeCoCu nanowire with the diameter modulated between 100 and 140 nm. As seen in the MFM image of **Figure 20(c)**, the single domain configuration of the nanowire is maintained independent of the number of modulations. Stray fields are emanating not only from the ends of the nanowire, as in the case of nanowires with constant diameter (**Figure 11(b,c)**) but also from the ends of the segments with larger diameters [29, 88].

Modulations of the magnetization profile lead to the creation of pinning sites for domain walls [37, 87]. Magnetic poles of different signs are created at the modulation transition regions,

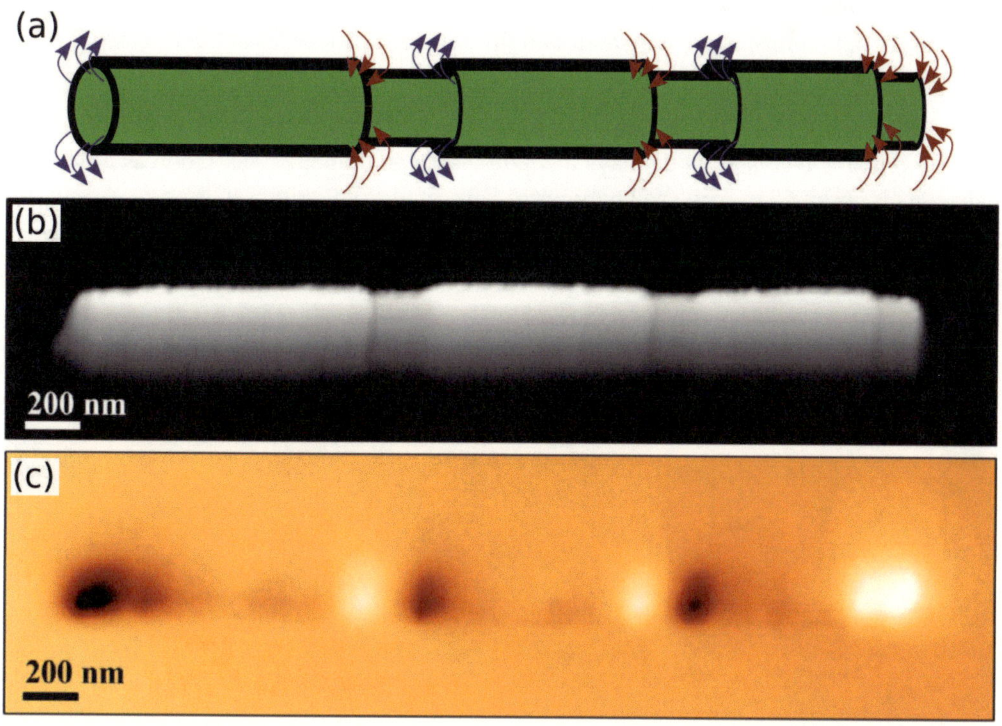

Figure 20. (a) Schematic of a diameter-modulated nanowire. (b) AFM image and (c) corresponding MFM image of a diameter-modulated FeCoCu nanowire (reprinted with permission from Rodríguez et al. [87]. Copyright 2016 American Chemical Society).

increasing the magnetic surface charge, which is minimized by the creation of vortex states. The efficiency of domain wall pinning sites in cylindrical nanowires depends on the parameters that modify the magnetization profile with respect to the magnetic surface charges. Hence, the pinning sites' geometrical magnetizations can be used to tailor the properties of cylindrical nanowires and integrate different functionalities into them.

7. Characterization of core-shell Fe-Fe_3O_4 nanowires

Energy-filtered transmission electron microscopy (EF-TEM) can be used to study the evolution of the core-shell nanowire structure for different oxidation times. TEM studies into core-shell Fe-Fe_3O_4 nanowires have revealed that annealing of both polycrystalline or single-crystal iron nanowires leads to the formation of an oxide shell structure (**Figure 21(a)**) but possess distinctly different characteristics [50]. The oxide shell thickness of polycrystalline Fe nanowires (**Figure 21(b,c)**) increases linearly with annealing time until the disappearance of the Fe core, leaving behind a pure Fe_3O_4 nanowire. However, the oxidation of the single-crystal Fe nanowires (**Figure 21(d,e)**) is observed to be a slower process, and the shell thickness does not increase beyond a certain limit (12 nm), thus maintaining the Fe core structure. It was observed that the O clearly concentrates at the edges while in fully oxidized nanowires; O is distributed across the entire volume. This can be attributed to the absence of grain boundaries that are required for heat-assisted oxygen diffusion. The ability of single-crystal core-shell nanowires to resist complete oxidation is a great advantage for applications that require high-temperature operations, while they maintain the magnetic properties of the Fe core [50].

Magnetic force microscopy studies into polycrystalline and single-crystal core-shell Fe-Fe_3O_4 nanowires reveal marked differences. The single-crystal Fe and the core-shell nanowires fabricated from single-crystal Fe displays a single magnetic domain state, whereas a multidomain state is observed in core-shell nanowires fabricated from polycrystalline Fe nanowires.

The core-shell nanowire is an attractive candidate for biomedical applications [7, 8, 83], due to their tunable magnetization and long-term stability. In addition to this, a high degree of

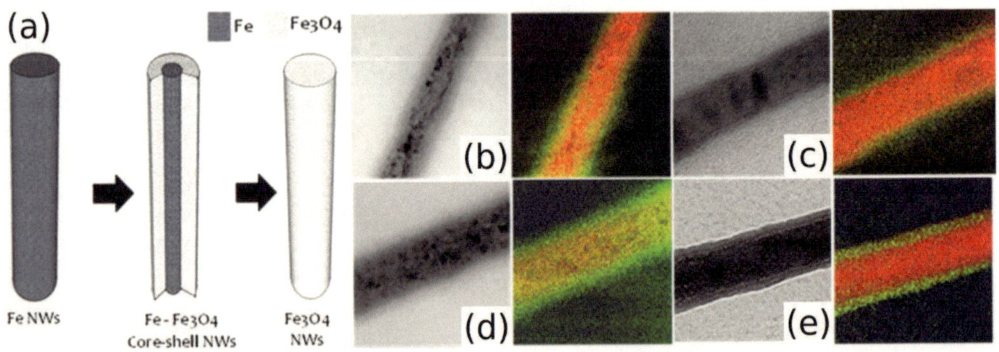

Figure 21. (a) Formation of the Fe-Fe_3O_4 core-shell structure, (b–e) bright field color-coded EF-TEM images of the Fe (red) and O (green) of polycrystalline Fe nanowires after (b) 20 min, (c) 24 h annealing and single crystalline Fe nanowires after (d) 1 h, (e) 72 h [50].

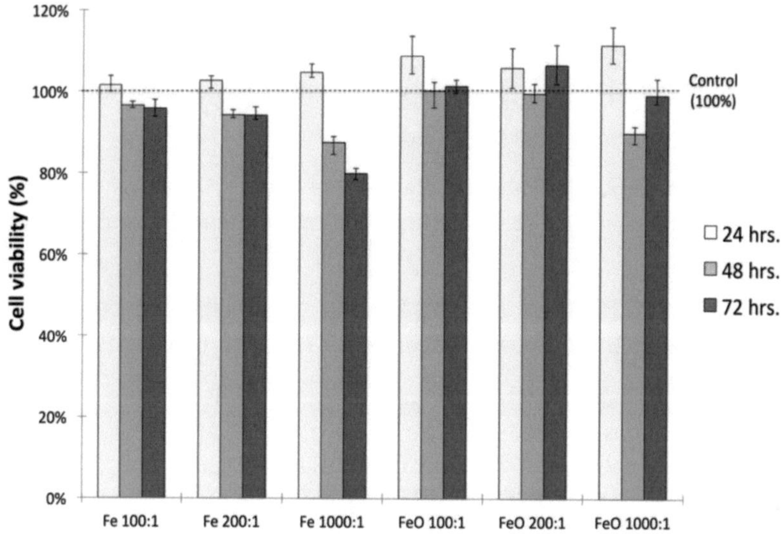

Figure 22. Cell viability studies of colon cancer cells incubated with Fe and Fe-Fe_3O_4 core-shell nanowires with varying concentrations (nanowire-to- cell ratio) for 24, 48, and 72 h [50].

biocompatibility required for biomedical applications has been demonstrated by these nanowires. For instance, studies into the metabolic activity of cells incubated with Fe and Fe-Fe_3O_4 core-shell nanowires of varying concentrations and incubation times (**Figure 22**) reveal that the nanowires have comparable and high values of cell viability at concentration of 100 and 200 nanowires per cell. At very high concentrations of 1000 nanowires per cell, Fe nanowires cause a decrease in cell viability to about 80% after 72 h, while viability with Fe-Fe_3O_4 core-shell nanowires remains high.

Author details

Hanan Mohammed, Julian A. Moreno and Jürgen Kosel*

*Address all correspondence to: jurgen.kosel@kaust.edu.sa

King Abdullah University of Science and Technology, Thuwal, Saudi Arabia

References

[1] Alfadhel A, Li B, Zaher A, Yassine O, Kosel J. A magnetic nanocomposite for biomimetic flow sensing. Lab on a Chip. 2014;**14**:4362-4369

[2] Fratila RM, Rivera-Fernández S, Jesús M. Shape matters: Synthesis and biomedical applications of high aspect ratio magnetic nanomaterials. Nanoscale. 2015;**7**:8233-8260

[3] Alfadhel A, Kosel J. Magnetic nanocomposite cilia tactile sensor. Advanced Materials. 2015;**27**:7888-7892

[4] Alnassar M, Alfadhel A, Ivanov YP, Kosel J. Magnetoelectric polymer nanocomposite for flexible electronics. Journal of Applied Physics. 2015;**117**:17D711

[5] Contreras MF, Sougrat R, Zaher A, Ravasi T, Kosel J. Non-chemotoxic induction of cancer cell death using magnetic nanowires. International Journal of Nanomedicine. 2015;**10**:2141

[6] Yassine O, Zaher A, Li EQ, Alfadhel A, Perez JE, Kavaldzhiev M, et al. Highly efficient thermoresponsive nanocomposite for controlled release applications. Scientific Reports. 2016;**6**:28539

[7] Martínez-Banderas AI, Aires A, Teran FJ, Perez JE, Cadenas JF, Alsharif N, et al. Functionalized magnetic nanowires for chemical and magneto-mechanical induction of cancer cell death. Scientific Reports. 2016;**6**:35786

[8] Shore D, Pailloux SL, Zhang J, Gage T, Flannigan DJ, Garwood M, et al. Electrodeposited Fe and Fe-Au nanowires as MRI contrast agents. Chemical Communications. 2016; **52**:12634-12637

[9] García-Martín J, Thiaville A, Miltat J, Okuno T, Vila L, Piraux L. Imaging magnetic vortices by magnetic force microscopy: Experiments and modelling. Journal of Physics D: Applied Physics. 2004;**37**:965

[10] Piraux L, George J, Despres J, Leroy C, Ferain E, Legras R, et al. Giant magnetoresistance in magnetic multilayered nanowires. Applied Physics Letters. 1994;**65**:2484-2486

[11] Piraux L, Renard K, Guillemet R, Mátéfi-Tempfli S, Mátéfi-Tempfli M, Antohe VA, et al. Template-grown NiFe/Cu/NiFe nanowires for spin transfer devices. Nano Letters. 2007;**7**:2563-2567

[12] Wang Z, Kuok M, Ng S, Lockwood D, Cottam M, Nielsch K, et al. Spin-wave quantization in ferromagnetic nickel nanowires. Physical Review Letters. 2002;**89**:027201

[13] Wernsdorfer W, Hasselbach K, Benoit A, Barbara B, Doudin B, Meier J, et al. Measurements of magnetization switching in individual nickel nanowires. Physical Review B. 1997;**55**:11552

[14] Kou X, Fan X, Dumas RK, Lu Q, Zhang Y, Zhu H, et al. Memory effect in magnetic nanowire arrays. Advanced Materials. 2011;**23**:1393-1397

[15] Mohammed H, Vidal EV, Ivanov YP, Kosel J. Magnetotransport measurements of domain wall propagation in individual multisegmented cylindrical nanowires. IEEE Transactions on Magnetics. 2016;**52**:1-5

[16] Whitney T, Jiang J, Searson P, Chien C. Fabrication and magnetic properties of arrays of metallic nanowires. Science. 1993;**261**:1316-1320

[17] Metzger RM, Konovalov VV, Sun M, Xu T, Zangari G, Xu B, et al. Magnetic nanowires in hexagonally ordered pores of alumina. IEEE Transactions on Magnetics. 2000;**36**:30-35

[18] Nielsch K, Müller F, Li A-P, Gösele U. Uniform nickel deposition into ordered alumina pores by pulsed electrodeposition. Advanced Materials. 2000;**12**:582-586

[19] Pitzschel K, Bachmann J, Martens S, Montero-Moreno JM, Kimling J, Meier G, et al. Magnetic reversal of cylindrical nickel nanowires with modulated diameters. Journal of Applied Physics. 2011;**109**:033907

[20] Sulka GD, Brzózka A, Liu L. Fabrication of diameter-modulated and ultrathin porous nanowires in anodic aluminum oxide templates. Electrochimica Acta. 2011;**56**:4972-4979

[21] Masuda H, Fukuda K. Ordered metal nanohole arrays made by a two-step replication of honeycomb structures of anodic alumina. Science. 1995;**268**:1466

[22] Lee W, Park S-J. Porous anodic aluminum oxide: Anodization and templated synthesis of functional nanostructures. Chemical Reviews. 2014;**114**:7487-7556

[23] Nielsch K, Hertel R, Wehrspohn R, Barthel J, Kirschner J, Gosele U, et al. Switching behavior of single nanowires inside dense nickel nanowire arrays. IEEE Transactions on Magnetics. 2002;**38**:2571-2573

[24] Sulka GD. Highly ordered anodic porous alumina formation by self-organized anodizing. Nanostructured Materials in Electrochemistry. 2008;**1**:1-116

[25] Kartopu G, Yalçın O. Fabrication and applications of metal nanowire arrays electrodeposited in ordered porous templates. In: Electrodeposited Nanowires and Their Applications. InTech; 2010

[26] Jaafar M, Navas D, Hernández-Vélez M, Baldonedo J, Vázquez M, Asenjo A. Nanoporous alumina membrane prepared by nanoindentation and anodic oxidation. Surface Science. 2009;**603**:3155-3159

[27] Masuda H, Kanezawa K, Nishio K. Fabrication of ideally ordered nanohole arrays in anodic porous alumina based on nanoindentation using scanning probe microscope. Chemistry Letters. 2002;**31**:1218-1219

[28] Liu C, Datta A, Wang Y. Ordered anodic alumina nanochannels on focused-ion-beam-prepatterned aluminum surfaces. Applied Physics Letters. 2001;**78**:120-122

[29] Choi J, Nielsch K, Reiche M, Wehrspohn R, Gösele U. Fabrication of monodomain alumina pore arrays with an interpore distance smaller than the lattice constant of the imprint stamp. Journal of Vacuum Science & Technology, B: Microelectronics and Nanometer Structures: Processing, Measurement, and Phenomena. 2003;**21**:763-766

[30] Maqableh MM, Tan L, Huang X, Cobian R, Norby G, Victora R, et al. CPP GMR through nanowires. IEEE Transactions on Magnetics. 2012;**48**:1744-1750

[31] Sung S-Y, Maqableh MM, Huang X, Reddy KSM, Victora R, Stadler BJ. Metallic 10 nm diameter magnetic sensors and large-scale ordered arrays. IEEE Transactions on Magnetics. 2014;**50**:1-5

[32] Iglesias-Freire Ó, Bran C, Berganza E, Mínguez-Bacho I, Magén C, Vázquez M, et al. Spin configuration in isolated FeCoCu nanowires modulated in diameter. Nanotechnology. 2015;**26**:395702

[33] Minguez-Bacho I, Rodriguez-López S, Vázquez M, Hernández-Vélez M, Nielsch K. Electrochemical synthesis and magnetic characterization of periodically modulated Co nanowires. Nanotechnology. 2014;**25**:145301

[34] Woo L, Ji R, Gösele U, Nielsch K. Fast fabrication of long-range ordered porous alumina membranes by hard anodization. Nature Materials. 2006;**5**:741

[35] Parkhutik V, Shershulsky V. Theoretical modelling of porous oxide growth on aluminium. Journal of Physics D: Applied Physics. 1992;**25**:1258

[36] Lee W, Schwirn K, Steinhart M, Pippel E, Scholz R, Gösele U. Structural engineering of nanoporous anodic aluminium oxide by pulse anodization of aluminium. Nature Nanotechnology. 2008;**3**:234-239

[37] Losic D, Lillo M. Porous alumina with shaped pore geometries and complex pore architectures fabricated by cyclic anodization. Small. 2009;**5**:1392-1397

[38] Salem MS, Sergelius P, Corona RM, Escrig J, Görlitz D, Nielsch K. Magnetic properties of cylindrical diameter modulated Ni80Fe20 nanowires: Interaction and coercive fields. Nanoscale. 2013;**5**:3941-3947

[39] Mohler JB, Sedusky HJ. Electroplating for the Metallurgist, Engineer, and Chemist. Chemical Publishing Company; New York. 1951

[40] Sousa C, Leitao D, Proenca M, Apolinario A, Correia J, Ventura J, et al. Tunning pore filling of anodic alumina templates by accurate control of the bottom barrier layer thickness. Nanotechnology. 2011;**22**:315602

[41] Vilanova Vidal E, Ivanov YP, Mohammed H, Kosel J. A detailed study of magnetization reversal in individual Ni nanowires. Applied Physics Letters. 2015;**106**:032403

[42] Ivanov YP, Chuvilin A, Lopatin S, Kosel J. Modulated magnetic nanowires for controlling domain wall motion: Toward 3D magnetic memories. ACS Nano. 2016;**10**:5326-5332

[43] Özkale B, Shamsudhin N, Chatzipirpiridis G, Hoop M, Gramm F, Chen X, et al. Multisegmented FeCo/Cu nanowires: electrosynthesis, characterization, and magnetic control of biomolecule desorption. ACS Applied Materials & Interfaces. 2015;**7**:7389-7396

[44] Sharma A, Zhu Y, Thor S, Zhou F, Stadler B, Hubel A. Magnetic barcode nanowires for osteosarcoma cell control, detection and separation. IEEE Transactions on Magnetics. 2013;**49**:453-456

[45] Liu K, Nagodawithana K, Searson P, Chien C. Perpendicular giant magnetoresistance of multilayered Co/Cu nanowires. Physical Review B. 1995;**51**:7381

[46] Yahalom J, Tessier D, Timsit R, Rosenfeld A, Mitchell D, Robinson P. Structure of composition-modulated Cu/Ni thin films prepared by electrodeposition. Journal of Materials Research. 1989;**4**:755-758

[47] Palmero EM, Béron F, Bran C, del Real RP, Vázquez M. Magnetic interactions in compositionally modulated nanowire arrays. Nanotechnology. 2016;**27**:435705

[48] Moreno JA. Multi-Segmented Magnetic Nanowires Fabrication and Characterization. King Abdullah University of Science and Technology. Thuwal, Makkah, Saudi Arabia. 2016

[49] Armelles G, Cebollada A, García-Martín A, Montero-Moreno J, Waleczek M, Nielsch K. Magneto-optical properties of core–shell magneto-plasmonic Au-$Co_xFe_{3-x}O_4$ nanowires. Langmuir. 2012;**28**:9127-9130

[50] Ivanov YP, Alfadhel A, Alnassar M, Perez JE, Vazquez M, Chuvilin A, et al. Tunable magnetic nanowires for biomedical and harsh environment applications. Scientific Reports. 2016;**6**:24189

[51] Ovejero JG, Bran C, Vilanova E, Kosel J, Morales MP, Vazquez M. Electrochemical synthesis of core-shell magnetic nanowires. Journal of Magnetism and Magnetic Materials. 2015;**389**:144-147

[52] Ozel T, Bourret GR, Mirkin CA. Coaxial lithography. Nature Nanotechnology. 2015 May 1;**10**(4):319-24

[53] Mayer B, Rudolph D, Schnell J, Morkötter S, Winnerl J, Treu J, et al. Lasing from individual GaAs-AlGaAs core-shell nanowires up to room temperature. Nature Communications. 2013;**4**:2931

[54] Vivas L, Ivanov YP, Trabada D, Proenca M, Chubykalo-Fesenko O, Vázquez M. Magnetic properties of Co nanopillar arrays prepared from alumina templates. Nanotechnology. 2013;**24**:105703

[55] Asenjo A, Jaafar M, Navas D, Vázquez M. Quantitative magnetic force microscopy analysis of the magnetization process in nanowire arrays. Journal of Applied Physics. 2006;**100**:023909

[56] Dubois S, Colin J, Duvail J, Piraux L. Evidence for strong magnetoelastic effects in Ni nanowires embedded in polycarbonate membranes. Physical Review B. 2000;**61**:14315

[57] Li A, Müller F, Birner A, Nielsch K, Gösele U. Polycrystalline nanopore arrays with hexagonal ordering on aluminum. Journal of Vacuum Science & Technology, A: Vacuum, Surfaces, and Films. 1999;**17**:1428-1431

[58] Williams DB, Carter CB, Veyssiere P. Transmission Electron Microscopy: A Textbook for Materials Science. Vol. 10. Springer. New York; 1998

[59] Zhang J, Ma H, Zhang S, Zhang H, Deng X, Lan Q, et al. Nanoscale characterisation and magnetic properties of Co81Cu19/Cu multilayer nanowires. Journal of Materials Chemistry C. 2015;**3**:85-93

[60] Egerton RF. Electron Energy-Loss Spectroscopy in the Electron Microscope. Springer Science & Business Media; 2011

[61] Pignard S, Goglio G, Radulescu A, Piraux L, Dubois S, Declemy A, et al. Study of the magnetization reversal in individual nickel nanowires. Journal of Applied Physics. 2000;**87**:824-829

[62] Tiwari U, Ghosh R, Sen P. Theory of magneto-optic Kerr effects. Physical Review B. 1994;**49**:2159

[63] Hertel R, Kirschner J. Magnetization reversal dynamics in nickel nanowires. Physica B: Condensed Matter. 2004;**343**:206-210

[64] Rugar D, Mamin H, Guethner P, Lambert S, Stern J, McFadyen I, et al. Magnetic force microscopy: General principles and application to longitudinal recording media. Journal of Applied Physics. 1990;**68**:1169-1183

[65] Lopatin S, Ivanov YP, Kosel J, Chuvilin A. Multiscale differential phase contrast analysis with a unitary detector. Ultramicroscopy. 2016;**162**:74-81

[66] Ivanov YP, Lopatin S, Kosel J, Chuvilin A. Unitary detector DPC imaging with multi-scale capabilities for analysis of local magnetic field of nanomaterials. Microscopy and Microanalysis. 2016;**22**:1704-1705

[67] Da Col S, Jamet S, Rougemaille N, Locatelli A, Mentes TO, Burgos BS, et al. Observation of Bloch-point domain walls in cylindrical magnetic nanowires. Physical Review B. 2014;**89**:180405

[68] Kimling J, Kronast F, Martens S, Böhnert T, Martens M, Herrero-Albillos J, et al. Photoemission electron microscopy of three-dimensional magnetization configurations in core-shell nanostructures. Physical Review B. 2011;**84**:174406

[69] Akhtari-Zavareh A, Carignan L, Yelon A, Ménard D, Kasama T, Herring R, et al. Off-axis electron holography of ferromagnetic multilayer nanowires. Journal of Applied Physics. 2014;**116**:023902

[70] Scholz W, Fidler J, Schrefl T, Suess D, Forster H, Tsiantos V. Scalable parallel micro-magnetic solvers for magnetic nanostructures. Computational Materials Science. 2003;**28**:366-383

[71] Fischbacher T, Franchin M, Bordignon G, Fangohr H. A systematic approach to multiphysics extensions of finite-element-based micromagnetic simulations: Nmag. IEEE Transactions on Magnetics. 2007;**43**:2896-2898

[72] Donahue M, Porter D. The Object Oriented Micromagnetic Framework (OOMMF) Project at ITL/NIST The OOMMF code is available at http://math.nist.gov/oommf 1998

[73] Scheinfein, M. R. LLG Micromagnetic SimulatorTM (http://llgmicro.home.mindspring.com)

[74] Tannous C, Gieraltowski J. The Stoner-Wohlfarth model of ferromagnetism. European Journal of Physics. 2008;**29**:475

[75] Fusil S, Piraux L, Mátéfi-Tempfli S, Mátéfi-Tempfli M, Michotte S, Saul C, et al. Nanolithography based contacting method for electrical measurements on single template synthesized nanowires. Nanotechnology. 2005;**16**:2936

[76] Fert A, Piraux L. Magnetic nanowires. Journal of Magnetism and Magnetic Materials. 1999;**200**:338-358

[77] Vila L, Piraux L, George J, Faini G. Multiprobe magnetoresistance measurements on isolated magnetic nanowires. Applied Physics Letters. 2002;**80**:3805-3807

[78] Toimil Molares ME, Höhberger EM, Schaeflein C, Blick RH, Neumann R, Trautmann C. Electrical characterization of electrochemically grown single copper nanowires. Applied Physics Letters. 2003;**82**:2139-2141

[79] Tanase M, Silevitch D, Chien C, Reich D. Magnetotransport properties of bent ferromagnetic nanowires. Journal of Applied Physics. 2003;**93**:7616-7618

[80] Mohammed H, Corte-León H, Ivanov Y, Moreno J, Kazakova O, Kosel J. Angular magnetoresistance of nanowires with alternating cobalt and nickel segments. IEEE Transactions on Magnetics. 2017

[81] Wegrowe J, Kelly D, Franck A, Gilbert S, Ansermet J-P. Magnetoresistance of ferromagnetic nanowires. Physical Review Letters. 1999;**82**:3681

[82] McGuire T, Potter R. Anisotropic magnetoresistance in ferromagnetic 3D alloys. IEEE Transactions on Magnetics. 1975;**11**:1018-1038

[83] Ivanov YP, Chuvilin A, Lopatin S, Mohammed H, Kosel J. Direct observation of current-induced motion of a 3D vortex domain wall in cylindrical nanowires. ACS Applied Materials & Interfaces. 2017;**9**(20):16741-16744. DOI: 10.1021/acsami.7b03404

[84] Tejo F, Vidal-Silva N, Espejo A, Escrig J. Angular dependence of the magnetic properties of cylindrical diameter modulated Ni80Fe20 nanowires. Journal of Applied Physics. 2014;**115**:17D136

[85] Wieser R, Nowak U, Usadel K-D. Domain wall mobility in nanowires: Transverse versus vortex walls. Physical Review B. 2004;**69**:064401

[86] Pitzschel K, Moreno JMM, Escrig J, Albrecht O, Nielsch K, Bachmann J. Controlled introduction of diameter modulations in arrayed magnetic iron oxide nanotubes. ACS Nano. 2009;**3**:3463-3468

[87] Rodríguez LA, Bran C, Reyes D, Berganza E, Vázquez M, Gatel C, et al. Quantitative nanoscale magnetic study of isolated diameter-modulated FeCoCu nanowires. ACS Nano. 2016;**10**:9669-9678

[88] Berganza E, Bran C, Jaafar M, Vázquez M, Asenjo A. Domain wall pinning in FeCoCu bamboo-like nanowires. Scientific Reports. 2016;**6**:29702

Chapter 3

Nanoscale Magnetic Domain Memory

Karine Chesnel

Additional information is available at the end of the chapter

http://dx.doi.org/10.5772/intechopen.71076

"Magnets are a bit like humans.
Not only they attract each other, they also have the capability to remember."

Abstract

Magnetic domain memory (MDM) is the ability exhibited by certain magnetic materials to reproduce the exact same nanoscale magnetic domain pattern, even after it has been completely erased by an external magnetic field. In this chapter, we review the various circumstances under which this unusual phenomenon occurs. We explain how partial MDM was first observed in rough Co/Pt multilayers with perpendicular magnetization as a result of structural defects. We then show how 100 % MDM was achieved, even in smooth ferromagnetic films, by coupling Co/Pd multilayers to an antiferromagnetic IrMn template via exchange interactions. We describe how high MDM, extending throughout nearly the entirety of the magnetization process, is obtained when zero-field-cooling the material below its blocking temperature where exchange couplings occur. We also review the persistence of MDM through field cycling and while warming the material all the way up to the blocking temperature. Additionally, we discuss the spatial dependence of MDM, highlighting intriguing oscillatory behaviors suggesting magnetic correlations and rotational symmetries at the nanoscopic scales. Finally, we review the dependence of MDM on cooling conditions, revealing how MDM can be fully controlled, turned on and off, by adjusting the magnitude of the cooling field.

Keywords: magnetic domains, ferromagnetic films, perpendicular magnetic anisotropy, magnetic domain memory, defect-induced memory, exchange-coupling induced memory

1. Introduction to the principles of MDM

1.1. Magnetic domain patterns in ferromagnetic films

Ferromagnetic materials are typically composed of magnetic domains. A magnetic domain is a region where the magnetic moments carried by individual atoms, i.e., the atomic spins, align in the same direction due to exchange couplings, as illustrated in **Figure 1a**. Rather than forming one giant macroscopic magnetic domain, the material often breaks down into a multitude of microscopic domains of different orientations, as illustrated in **Figure 1b**. At the delimitation between one magnetic domain and a neighboring one, a coherent rotation of the atomic spins occurs. The delimitating region, called domain wall, may be small in respect to the domain size. Domain sizes typically range from 1 to 100 μm in bulk ferromagnetic materials, and from 100 nm to 1 μm in thin ferromagnetic films [1–3].

In a ferromagnetic material, the microscopic magnetic domains of various magnetization directions arrange in specific ways that minimize the competing magnetic energies present in the system. The three dimensional sum of the magnetic moments carried by the individual domains produces, at the macroscopic scale, a net magnetization ***M***. The magnitude and the direction of the net magnetization depends on how the magnetic moments of the individual domains are distributed throughout the material. While ***M*** may be uniquely set by applying an external magnetic field ***H***, following a specific magnetization procedure, the associated microscopic magnetic domain pattern, or topology, is usually not unique. The formation of magnetic domain topologies and their correlation with the magnetic history of the material is still today a vast field of research to be explored and understood.

Thin ferromagnetic films have been extensively studied during the past decades, in particular because of the variety of magnetic domain patterns they exhibit [4–8]. Depending on the composition, the crystallographic structure and the thickness of the film, the magnetization may point in-plane or out-of-plane. The magnetic domains may take various shapes, from round bubbles to elongated, almost infinite, stripes. Some thin film materials exhibit complex formations such as vortices [9–11] and skyrmions [12, 13], illustrated in **Figure 1c, d**, which have instigated a renewed interest in the recent years because of their potential applications in magnetic memory technologies and spintronics [14, 15].

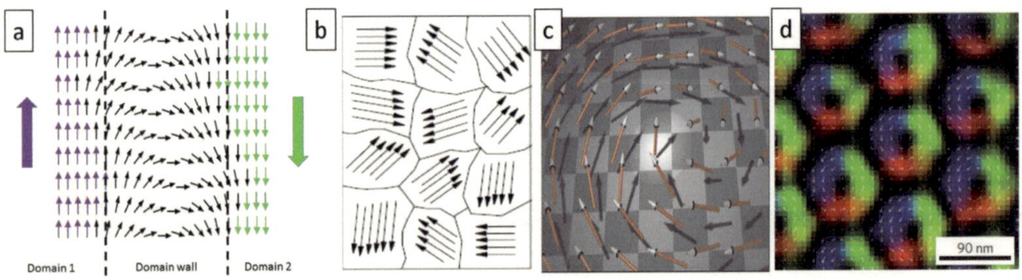

Figure 1. (a) Schematic illustrating the formation of magnetic domains by alignment of atomic spins. The schematic shows a domain wall, formed by the coherent rotation of spins. (b) Schematic of a collection of magnetic domains in a ferromagnetic material. (c) Schematic of a magnetic vortex core [10]. (d) Lorentz microscopy image of a skyrmion lattice in FeCoSi. Extracted from Fert et al. [14].

1.2. The case of thin films with perpendicular magnetic anisotropy

Thin ferromagnetic films with perpendicular magnetic anisotropy (PMA) [16] exhibit a particularly rich set of magnetic domain patterns. In PMA films, the domain magnetization points mostly out-of-plane, thus producing high demagnetization fields and leading to the formation of regular domain patterns and magnetic textures [17–21]. PMA films have attracted an accrued attention in the years 2000–2010s because the perpendicular magnetization has enabled significant domain size reductions, in comparison to the in-plane magnetic films, thus benefiting the magnetic recording industry. State-of-the-art ultra-high density magnetic recording technologies utilizes granular PMA thin film media, where magnetic domain sizes are as small as 20–50 nm [22].

In smooth PMA films, magnetic domains take a variety of shapes and they form patterns of various topologies. Such variety of shapes and topologies is for example observed in $[Co/Pt]_N$ multilayers where the thickness of Co typically varies between 5 and 50 Å, the thickness of the Pt layer is around 7 Å, and the number of repeat around N = 50. In these multilayers, the PMA is primarily achieved by exploiting the surface anisotropy created by the layering between the Co and the Pt layers, and the high number of repeats, as well as the magneto-crystalline anisotropy produced by the crystallographic texturing [23–26].

The magnetic domain topologies observed in these Co/Pt multilayers vary from bubble domain patterns to pure maze patterns formed of long interlaced striped domains, as shown in **Figure 2**. Both patterns in **Figure 2** are observed at remanence, where the external field is $\boldsymbol{H} = 0$. At that point, the net magnetization is $\boldsymbol{M} \approx 0$, so the area covered by the domains of a given direction nearly equals the area covered by the domains in the opposite direction. Both the bubble

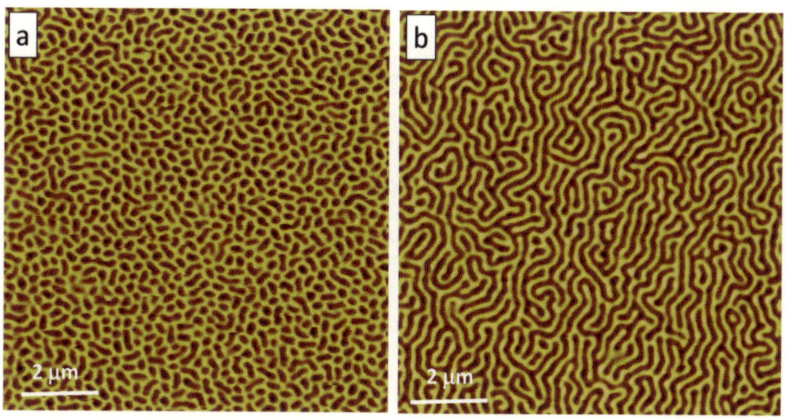

Figure 2. Magnetic Force Microscopy (MFM) images of magnetic domains in a $[Co(31\ Å)/Pt(7\ Å)]_{50}$ multilayer with perpendicular magnetic anisotropy. These domains patterns are measured at remanence (H = 0). The two colors represent opposite magnetization directions, pointing out-of-plane and into the plane. Image size is 10 × 10 µm. (a) Maze pattern forming after applying H = 2000 Oe; (b) bubble pattern forming after applying H = 9000 Oe. The saturation point is H = 11,400 Oe for this film. Extracted from Chesnel et al. [27].

pattern and the maze pattern, as well as a wide array of intermediate patterns, can be observed at remanence in the same multilayer. The topology of the remanent pattern mainly depends on the magnetic history of the film. In particular, it has been shown that the remanent magnetic topology drastically change when the magnitude H_m of the magnetic field applied perpendicular to the film changes. If the applied field is saturating, the film form a maze-like pattern at remanence. However, if H_m is set to a specific value slightly below the saturation point, a bubble pattern is achieved and the domain density is maximized [27]. Each time the external magnetic field is cycled, the magnetic domain pattern is erased. When the field is released back to $H = 0$, the new remanent domain pattern is, in this material, different from the previous one.

1.3. Magnetic domain memory

Magnetic domain memory (MDM) is the tendency for magnetic domains to retrieve the same exact same pattern after the pattern has been erased by a saturating magnetic field. When MDM occurs, not only the magnetic domains retrieve the same type of shape and size, but their distribution in space, or topology, is similar. MDM may be total, in which case the domain topology is identical; or MDM may be partial, in which case the domain pattern is similar, with some topological differences.

Ferromagnetic films usually do not exhibit any MDM. Like in the Co/Pt multilayers previously mentioned, the domain pattern at a given field point H generally does not repeat after the field has been cycled [28]. The magnetic domains may always take the form of small bubbles or of elongated stripes of specific size at that point in field, but their spatial distribution will be completely different after each cycle. This is well illustrated in **Figure 3**, which shows magnetic domain patterns in Co/Pt multilayers measured at remanence before and after applying a minor magnetization cycle. The initial pattern exhibits elongated striped domains

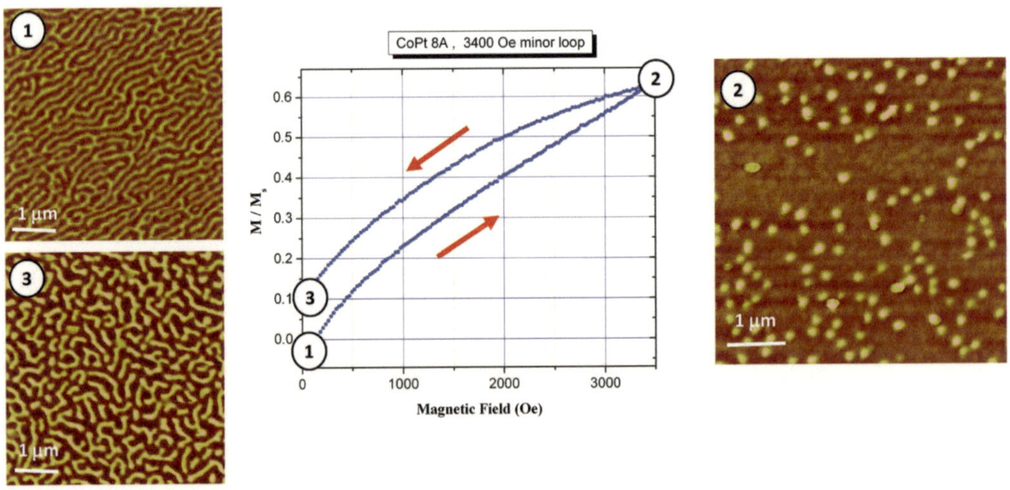

Figure 3. Evolution of the magnetic domain pattern in a $[Co(8\ Å)/Pt(7\ Å)]_{50}$ multilayer while cycling the magnetic field H through half a minor loop. Plotted is the net magnetization $M(H)$. The MFM images are 5 × 5 µm and are all collected at the same location of the film. (1) At remanence ($H = 0$), before applying the field; (2) at $H = 3400$ Oe, domains have shrunk down to small bubbles; (3) back to remanence ($H = 0$), after applying the field. Extracted from Westover et al. [28].

http://dx.doi.org/10.5772/intechopen.71076

of opposite up / down magnetization with close to a 50% up: 50% down coverage. At the highest magnetic field value, which is not quite saturating, the domains of reverse magnetization have shrunk down to small sparse bubbles. Upon return to zero field, the film exhibits magnetic stripes again but these stripes are shorter and their spatial distribution, or topology, is completely different than in the initial image. In this case, the film does not exhibit any MDM.

MDM has been so far only observed in specific thin ferromagnetic films that have specific structural or magnetic properties. As explained in the following sections, MDM has first been discovered in rough Co/Pt films, due to the presence of defects, acting as pinning sites for the domain nucleation. In that case, the observed defect-induced MDM is partial and only occurs in the nucleation phase of the magnetization cycle. We will see how MDM can however be maximized by exploiting magnetic exchange between the ferromagnetic (F) layer and an antiferromagnetic (AF) layer. This has been successfully achieved in [Co/Pd] IrMn films where F Co/Pd multilayers are sandwiched in between AF IrMn layers. The film is field-cooled (FC) down below its blocking temperature, to allow exchange couplings to occur. Under these conditions, the magnetic domain pattern imprinted in the IrMn layer plays the role of a magnetic template. In this case, the observed exchange-bias induced MDM reaches 100% throughout a large portion of the magnetization process.

2. Probing MDM

2.1. Real space imaging

Magnetic domain patterns may be directly measured via Magnetic Force Microscopy (MFM), as seen in **Figures 2** and **3**. The MFM technique allows the visualization of magnetic domains by probing the out-of-plane magnetic stray fields emanating from the surface of the film. With a spatial resolution down to about 20–25 nm, MFM allows the detection of individual magnetic domains in thin PMA ferromagnetic films, as these domains are typically 50–200 nm wide.

The investigation of MDM may be possible via MFM under certain experimental conditions. In 2003, Kappenberger et al. [29] showed in their study of CoO/[Co/Pt] multilayers via MFM that the specific magnetic domain pattern initially observed in a particular region of the film was fully recovered after magnetically saturating the material, as seen in **Figure 4**. In this experiment, the material was zero-field-cooled (ZFC) down to 7.5 K and then imaged at 7.5 K while applying a large magnetic field up to 7 T. The initial striped domain pattern observed in the Co/Pt layer immediately after ZFC was completely erased under application of magnetic field, but it was fully retrieved when the magnetic field was decreased back to near coercive point H_c on the descending branch of the magnetization loop. At H_c, the net magnetization is near zero, like it was during ZFC.

Such experiment requires very sophisticated MFM instrumentation, with cryogenic and in-situ magnetic field capabilities. Only very few MFM instruments in the world allow such extreme temperature and field environments. Because of the difficulty to measure MFM images at low temperature and under high in-situ magnetic field, the study of MDM via MFM is somewhat limited in practice. The magnetization of the MFM tip may reverse while

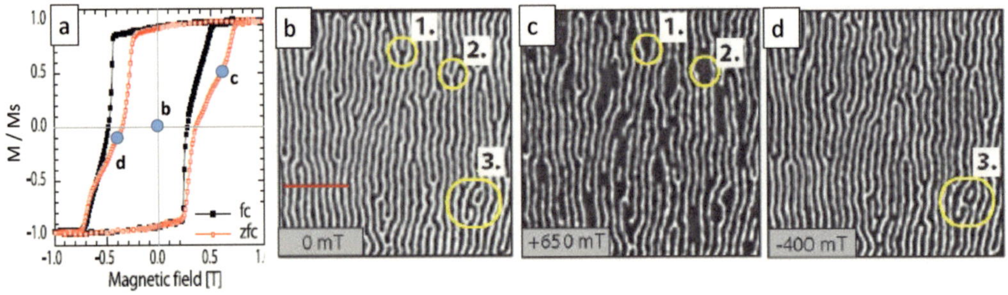

Figure 4. Evolution of the magnetic domain pattern in an exchange-biased CoO/ [Co/Pt] multilayer. The MFM images are 5 × 5 μm. (a) Magnetization loops *M(H)* measured perpendicular to the film at 5 K after field-cooling (FC) under 1 T and zero-field-cooling (ZFC) the film. (b) Initial striped domain pattern measured right after ZFC. (c) At H = +650 mT on the ascending branch of the magnetization loop, the white domains have shrunk or disappeared. (d) At H = −400 mT on the descending branch of the magnetization loop after saturating, the pattern is exactly reverse of the initial pattern in (a). Extracted from Kappenberger et al. [29].

the in-situ magnetic field is being varied. Also the maximum possible value for the magnetic field may be smaller than the value required to saturate the material, making the collection of MFM images throughout the entire magnetization loop impossible. Lastly, MFM images are typically micrometric, covering a few microns of the film, and a handful of magnetic domains, thus providing a localized view only.

2.2. Coherent x-ray resonant magnetic scattering

Complementary to MFM, the technique of coherent x-ray magnetic scattering [30, 31] is a powerful tool to study MDM. Under coherent illumination, the material produces a speckled scattering pattern that is a unique fingerprint of the charge and magnetic configuration of the material [32, 33], as illustrated in **Figure 5**. Because x-rays are insensitive to magnetic fields, an in-situ magnetic field of any value and any direction can be applied to the material while collecting the x-ray scattering signal. This allows the study of MDM throughout the entire magnetization process. Also, if mounted on a cryogenic holder, the material can be cooled down under various

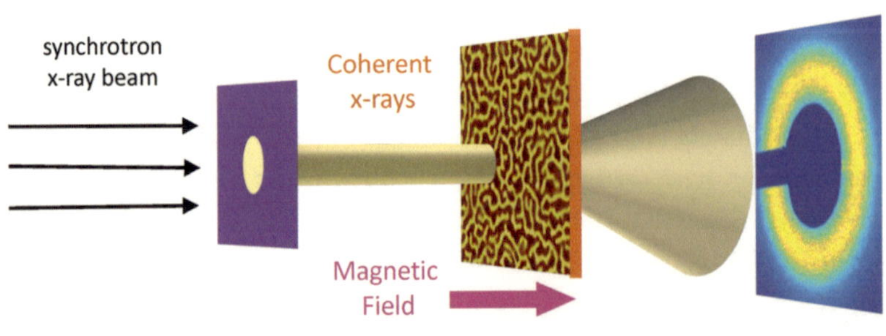

Figure 5. Layout for coherent x-ray resonant magnetic scattering (C-XRMS). The synchrotron x-ray beam is spatially filtered by a pinhole to enhance the transvers coherence. The energy of the x-ray is finely tuned to a magnetic resonance, thus providing longitudinal coherence and magneto-optical contrast. Under coherent illumination, the magnetic domains in the film produce a speckle pattern collected on a CCD camera. Extracted from Chesnel et al. [33].

field cooling conditions, and x-ray scattering can be measured at low temperature. Lastly, the region illuminated by the x-rays is set by the size of the spatial filter used to obtain transverse coherence. The illumination spot size, typically in the order of 20–50 μm, allows covering a relatively large number of magnetic domains, thus providing useful statistical information [34].

A magneto-optical contrast can be obtained by finely tuning the x-ray energy to specific resonance edges associated to the magnetic elements present in the material. This technique, called x-ray resonant magnetic scattering (XRMS), offers chemical selectivity [35–37]. Typical resonant edges used in XRMS are the L_2 and L_3 edges which correspond to the excitation of electrons from the (2p) to the (3d) electronic bands in transition metals. For iron (Fe), the $L_{2,3}$ edges are at around 708 and 720 eV, respectively. For cobalt (Co), the $L_{2,3}$ edges are at around 778 and 791 eV, respectively. At these energies (soft x-rays), the x-ray wavelength is around 1–2 nm, thus allowing the probing of magnetic structures in ferromagnetic films, with spatial resolutions down to few nanometers.

By collecting XRMS patterns at various magnetic field values, one can follow the evolution of the magnetic configuration of the material throughout the magnetization process. An example of such experiment is shown in **Figure 6**, where XRMS patterns collected on the descending branch of the magnetization loop are compared to MFM images at nearly same field values. At saturation, no magnetic domain exists so the XRMS pattern has no magnetic signal, only a pure charge background. Shortly after nucleation, sparse domains have nucleated and started to expand, leading to a diffuse disk around the center of the XRMS pattern. As the domain propagation progresses, interlaced domains of opposite directions fill the entire space, leading to a ring-like XRMS pattern. The radius of the ring relates to the magnetic period existing in the material at that stage [38].

Combining coherent x-rays with XRMS, the technique of *coherent*-XRMS (C-XRMS) turns out to be very useful to study MDM [33]. Measured in the scattering space, or the so-called reciprocal space, the CXRMS pattern provides an indirect view of magnetic domains. The inversion of the CXRMS pattern into the real space image is a complex process due to a phase loss in the intensity of the scattering signal [32]. However, if the x-ray beam is made coherent, the

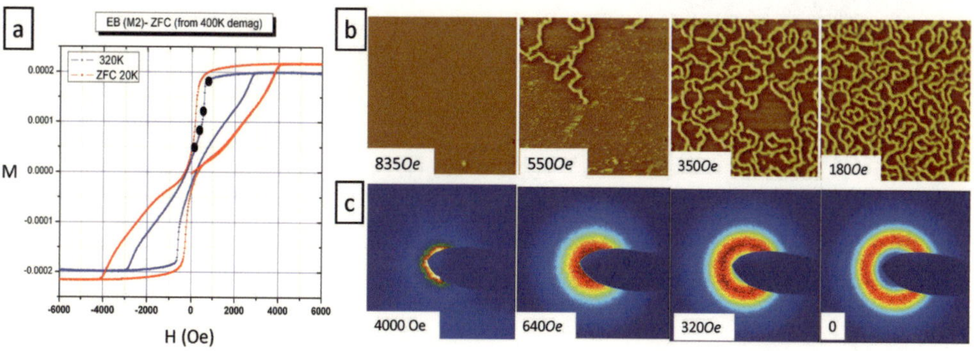

Figure 6. Probing magnetic changes in a [Co/Pd]IrMn film via MFM and XRMS. (a) Magnetization loops measured at 320 K and at 20 K after ZFC. (b) Evolution of the magnetic domain pattern at 300 K, viewed via 10 × 10 μm images. (c) Evolution of the XRMS pattern at 20 K in ZFC state at similar points in magnetization. Extracted from Chesnel et al. [38].

x-ray scattering pattern produced by the material will show speckles, as illustrated in **Figure 7**. The specific speckle pattern is a unique fingerprint of the magnetic domain pattern in the real space. Comparing speckle patterns collected at different points throughout magnetization then allows the evaluation of MDM in the material.

2.3. Speckle cross-correlation metrology

To evaluate MDM via CXRMS, speckle patterns collected at different field values are compared. This comparison is typically done via cross-correlation [39, 40]. In the cross-correlation process, the intensity in the two images is compared pixel by pixel. In practice, this done by multiplying the intensity of two image, pixel by pixel, while one image is spatially shifted in respect to the other one. Mathematically, this operation may be written as follows:

$$A \times B = \sum_{i,j} A(i,j)B(X+i,\ Y+j) \tag{1}$$

where A and B are the two images being correlated, $A(i, j)$ represents the intensity of image A at pixel (i,j) and $B(X+i,\ Y+j)$ represents the intensity of image B at pixel $(X+i,\ Y+j)$. Note that a shift of X pixels in the first direction and Y pixels in the second direction is applied between the two images. The sum takes care of covering all the pixels present in the images. The resulting cross-correlation product $A \times B$ is a function of the shift (X,Y). It forms a pattern

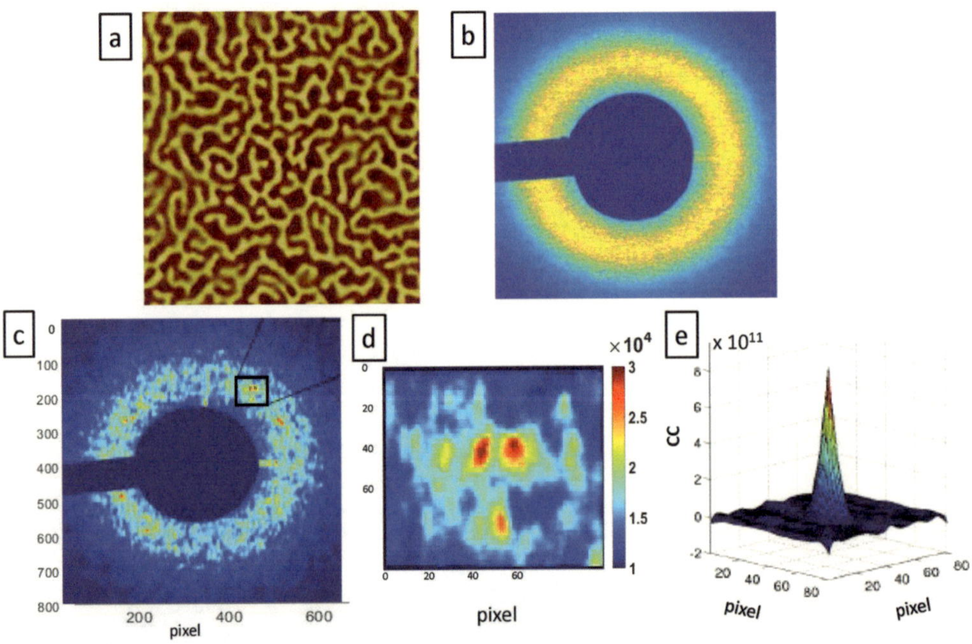

Figure 7. Extraction of speckle patterns and cross-correlation process. (a) Magnetic domain pattern in real space, measured via MFM. The image is 10 × 10μm. (b) Associated XRMS pattern as collected on the CCD detector. (c) Speckle pattern, or pure coherent component, extracted from the XRMS pattern (b). (d) Zoomed-in view on the speckle spots. (e) Correlation pattern showing a peak at its center. The area under the peak provides an estimate of the amount of correlation between two speckle patterns. Extracted from Chesnel et al. [33].

in the (X,Y) space, which is called correlation pattern. An example of such correlation pattern is shown in **Figure 7e**. To accelerate the cross-correlation operation, which may be computationally costly when comparing thousands of scattering images, the cross-correlation may be done via fast Fourier Transform (FFT) operations. When using FFT, the correlation pattern $A \times B$ is an image of same size than A and B. It generally shows a peak around its center, for which X = 0 and Y = 0. The intensity under the peak provides the amount of correlation between the two images. The width of the peak in the (X,Y) space corresponds to the average speckle size, which is generally set by the optics.

To quantify MDM, the intensity in the cross-correlation pattern $A \times B$ is integrated and compared to the intensity of the auto-correlated patterns $A \times A$ and $B \times B$. A normalized correlation coefficient is thus evaluated as follows:

$$\rho = \frac{\sum_{X,Y} A \times B}{\sqrt{\left(\sum_{X,Y} A \times A\right)\left(\sum_{X,Y} B \times B\right)}} \quad (2)$$

The coefficient ρ is then comprised between 0 and 1. If the two images A and B are completely different, ρ will be close to 0. If the two images A and B are exactly the same, $\rho = 1$ (or 100%). Since each of the correlated speckle patterns is a unique fingerprint of the magnetic domain configuration, the correlation coefficient ρ can be used to quantify MDM. When ρ is low, magnetic domain patterns are very different, and there is no or little MDM. When ρ is high (close to 1), magnetic domain patterns are very similar, and MDM is close to 100%.

2.4. Mapping MDM

MDM can be evaluated in a number of different ways. The most straightforward way to evaluate MDM is to compare magnetic speckle patterns measured at the same point H in magnetic field, after a full cycle has been completed, as illustrated in **Figure 8a**. This comparison is called returned point memory (RPM) [39]. An alternative way is to compare magnetic

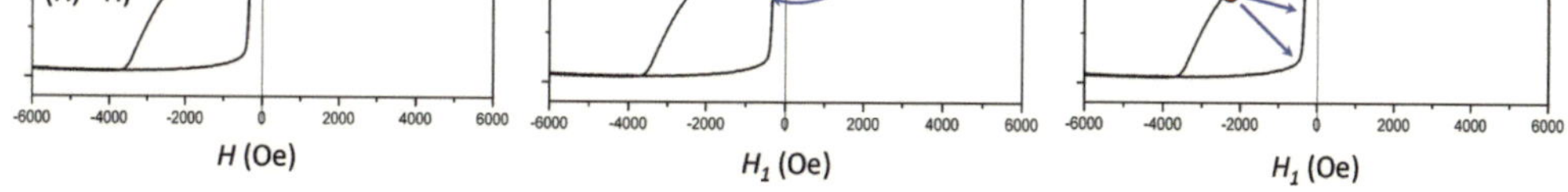

Figure 8. Illustration of various cross-correlation approaches. (a) Single point correlation. RPM: cross-correlating pairs of speckle patterns collected at same field (*H*, *H*). CPM: cross-correlating pairs of speckle patterns collected at opposite fields (*H*, −*H*). (b) Cross-field correlations AxA: correlating any point on the ascending branch with any other point on the ascending branch (c) cross-field correlations AXD: correlating any point on the ascending branch and any other point on the descending branch. In each approach, the two points to be correlated may be separated by several loops. Extracted from Chesnel et al. [38].

speckle patterns measured at opposite field values. In that case, one point is located at field H on the ascending branch of the magnetization loop whereas the other point is located at opposite field $-H$ is on descending branch. This comparison is called conjugate point memory (CPM). RPM and CPM can then be plotted as function of the magnetic field H, where H typically varies from negative saturation to positive saturation, on the ascending branch of the magnetization loop.

A more complete exploration of MDM may be performed by cross-correlating two speckle images A and B collected any field value H throughout the magnetization process [40]. The resulting correlation coefficient ρ may be then mapped in a two-dimensional field space (H_1, H_2) where H_1 is the field value of image A and H_2 is the field value of image B. The resulting $\rho(H_1, H_2)$ map is called correlation map. The two field values (H_1, H_2) may be both chosen on ascending branches, either on the same branch or on two different branches separated by one of more cycles. Such correlation map is denoted as A × A. A similar study can be done on the descending branch, leading to a D × D map. If the magnetization loop is symmetrical, the correlations on A × A and D × D are expected to be the same. In another approach, one field value H_1 may be chosen on the ascending branch and the other value H_2 on the descending branch, leading to an A × D map. Examples of such correlation maps $\rho(H_1, H_2)$ are shown in **Figure 13**.

More subtle measurements of MDM may exploit the spatial information contained in the CXRMS patterns. Instead of computing the cross-correlation on the entire image, the correlation is computed on specific portions of the image. One can for example look at the dependence on the scattering vector q, which is the distance from the center of the scattering pattern to a specific point in the image. The correlation is then performed on rings of specific q radii selected from the scattering patterns, as illustrated in **Figure 9**. The resulting correlation coefficient ρ may be plotted as a function of q, or mapped in a (q, H) space [40]. An example of such pattern is shown in **Figure 9d**. Other approaches explore the angular dependency of ρ.

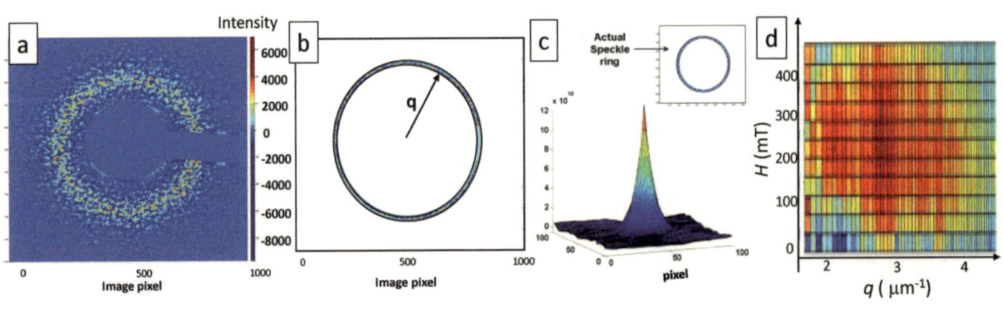

Figure 9 . Q-selective correlation process. (a) Initial speckle pattern; (b) ring selection from the speckle pattern; the radius of the ring is Q; (c) correlation pattern resulting from cross-correlating ring selections of two speckle patterns; (d) example of $\rho(Q, H)$ correlation map measure on a ZFC [co/Pd]/IrMn film. Extracted from Chesnel et al. [40].

3. Defect-induced MDM

3.1. First observations in Co/Pt thin films

The first studies of MDM via x-ray speckle correlation metrology were carried out on [Co/Pt] multilayered thin films by Pierce et al. [39] in 2003. These multilayers, made of 50 repeats of a Co/Pt bilayers, where the Co thickness is 0.4 nm, the Pt thickness is 0.7 nm, exhibit perpendicular anisotropy, leading to the formation of microscopic striped domain patterns.

These magnetic correlation studies demonstrated the occurrence of microscopic magnetic return-point-memory (RPM) when these films exhibited some interfacial roughness. It was found that smooth films with no roughness did not produce any RPM. However, films with large roughness produced significant RPM in the nucleation region of the magnetization process. It was established that the observed microscopic magnetic memory was induced by the presence of defects in the film, playing the role of anchors for magnetic domains to nucleate. This phenomenon is referred to as 'defect-induced' or 'disorder-induced' MDM [41, 42].

In **Figure 10**, magnetic domain configurations are shown for different film roughnesses. When the film is smooth (grown under 3 mT of Ar pressure), a typical interlaced stripe domain pattern occurs. When the film is rough (grown under 12 mT of Ar pressure), the magnetic domain pattern becomes fuzzy and the magnetic periodicity is somewhat lost. The associated XRMS scattering profile shows a well-defined peak for the 3 mT film, but a weaken peak for the 12 mT film.

In these studies, microscopic magnetic correlations were evaluated while cycling the magnetic field throughout minor loops and major loops. In particular RPM and CPM were measured at various field values along the ascending and the descending branches of the major magnetization loop. These measurements were carried out on films with various roughnesses.

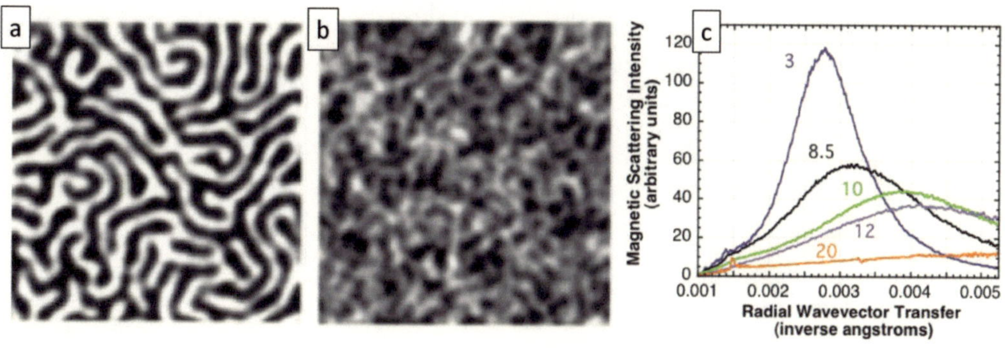

Figure 10. Effect of roughness on the magnetic domain configuration in [Co(4 Å)/Pt(7 Å)]$_{50}$ multilayers. (a) and (b) 3 x 3 μm MFM images for films grown under different Ar pressure (a) 3 mT, (b) 12 mT. (c) XRMS profiles for films with different roughnesses (the label indicates the Ar pressure in mT). Extracted from Pierce et al. [41].

3.2. RPM and CPM dependence on magnetic field

RPM and CPM were evaluated on the $[Co(0.4\ nm)/Pt(0.7\ nm)]_{50}$ multilayers at different points throughout the ascending and descending branches of the major loop after cycling the field multiple times. Co/Pt multilayers with significant interfacial roughness exhibited non-zero RPM and CPM, as illustrated in **Figure 11a**. It was found that CPM was systematically lower than RPM. This suggested that disorder has a component that breaks spin reversal symmetry. It was also found that both RPM and CPM exhibited the same trend: their value was larger at nucleation ($H = -1$ kOe) and decreased monotonically down to near zero when the magnetic field was increased toward saturation ($H = +4$ kOe). The high correlation value ρ at nucleation suggested that magnetic domains tend to nucleate at specific locations, which are pinned by the defects. The decreasing trend down to zero correlation when field is increasing suggests that once nucleated, the magnetic domains propagate rather randomly, or non-deterministically, throughout the film, leading to decorrelation [43].

3.3. Dependence on roughness

As expected, the observed defect-induced MDM depended on the amount of interfacial roughness exhibited by the $[Co(0.4\ nm)/Pt(0.7\ nm)]_{50}$ multilayers. If the film was relatively smooth, no memory was found. If, on the contrary, the film was rough enough, some RPM and CPM would be detected. The dependence of MDM with interfacial roughness is illustrated in **Figure 11b**, where RPM and CPM are measured at the coercive point. For roughness below 0.5 nm or so, no RPM and no CPM was observed. When the roughness exceeded 0.5 nm, both RPM and CPM quickly increased and started to plateau at around 0.7 nm of roughness, reaching about 40% for CPM and 50% for RPM. This behavior, measured at room temperature was, to some extent, reproduced by nonzero-temperature numerical simulations based on Ising models [41, 42].

The observation of microscopic magnetic memory induced by defects in [Co/Pt] multilayered thin films opened the door to new perspectives and led to further questions, both on the fundamental and applied levels. In particular, the observed CPM and RPM were somewhat limited in magnitude, not exceeding 50–60% in best cases, and also they occurred in a limited region of the magnetization process, namely the nucleation region.

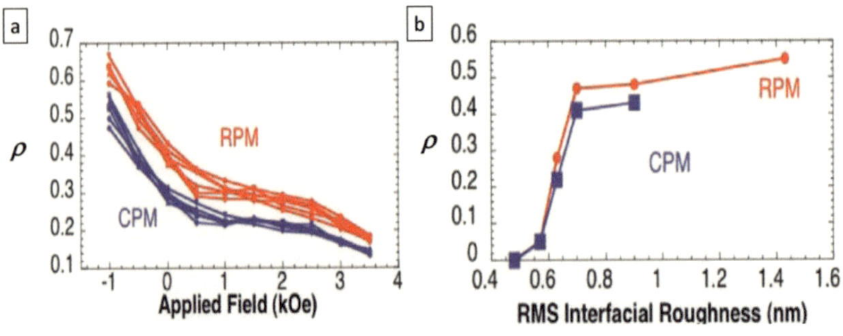

Figure 11. RPM and CPM correlations in rough $[Co(4\ Å)/Pt(7\ Å)]_{50}$ multilayers. (a) RPM and CPM vs. field on the ascending branch of the magnetization loop for a rough film grown under 8.5 mT Ar pressure. (b) RPM and CPM values at the coercive point, plotted as a function of interfacial roughness. Extracted from Pierce et al. [41].

http://dx.doi.org/10.5772/intechopen.71076

Would there be ways, in some materials, to increase MDM to higher values and to extend it to a wider region of the magnetization loop?

4. Exchange-bias induced MDM

In an attempt to increase the amount of MDM initially observed in rough ferromagnetic materials came the idea of incorporating exchange couplings in the film by interlaying the ferromagnetic (F) layer with an antiferromagnetic (AF) layer that would play the role of a magnetic template.

This was successfully achieved by combining a F Co/Pd multilayer with an AF IrMn layer [44]. After cooling the material below its blocking temperature T_B, and inducing exchange couplings (EC), high MDM was observed, which extended throughout almost the entire magnetization cycle. The observed MDM reached unprecedented values as high as 100%, even when the film was smooth.

If the material is cooled down under a non-zero field, the EC interactions produce a net exchange bias (EB) in the magnetization loop. However, because the origin of MDM is microscopic, high MDM can be observed even when zero-field cooling the material (in the absence of field), in which case not net bias exists.

4.1. First MDM observations in [Co/Pd]IrMn multilayers

The first observations of EC induced MDM were reported by Chesnel et al. [44] in 2008, in [Co/Pd]/IrMn multilayers. It was found that when zero-field-cooled (ZFC), the material exhibited high MDM. The magnetic correlations reached high values throughout a wide range of field values.

The material consisted of an interlay of F [Co(0.4 nm)/Pd (0.7 nm)]$_{12}$ multilayers with AF layers made of IrMn (2.4 nm) alloy, repeated 4 times. This film exhibited PMA, leading to the formation of serpentine magnetic domains that were about 150–200 nm wide, as seen in **Figure 12a**.

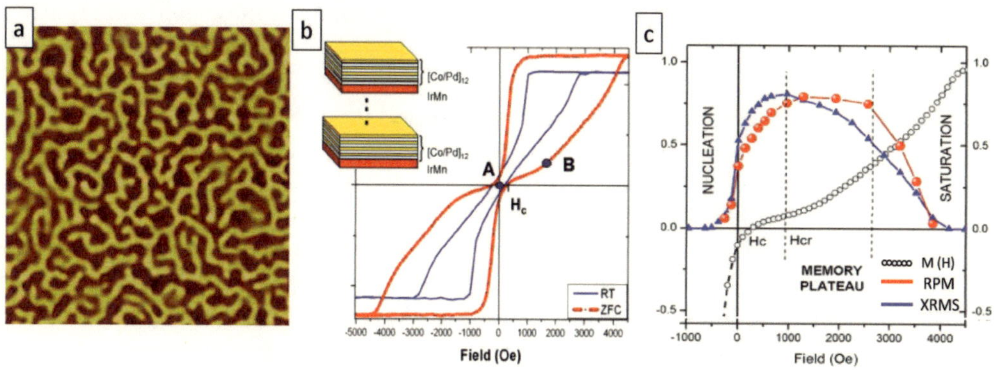

Figure 12. MDM measurements in ZFC [Co/Pd]/IrMn multilayers. (a) 10 × 10 μm MFM image of the striped magnetic domains. (b) Magnetization loops $M(H)$ at 300 K and at 20 K in ZFC state. (c) RPM vs. field H plotted against the ascending branch of the magnetization curve. Also plotted is the intensity of the XRMS signal. Extracted from Chesnel et al. [44].

To enhance the exchange couplings between the [Co/Pd] and the IrMn layers, the film was first demagnetized at high temperature and was then cooled down below its blocking temperature, $T_B \sim 300$ K, down to temperatures as low as 20 K. MDM was probed at 20 K using x-ray speckle correlation metrology. CXRMS patterns were collected at finely spaced field values throughout the magnetization loop seen in **Figure 12b** while an in-situ magnetic field was cycled numerous times.

4.2. Dependence on magnetic field throughout magnetization loop

The MDM observed in the ZFC [Co/Pd]/IrMn film shows an interesting behavior, which drastically differs from the behavior observed in rough Co/Pt films. **Figure 12c** shows the amount RPM measured in [Co/Pd]/IrMn at 20 K after ZFC and its dependence on magnetic field throughout the ascending branch magnetization loop. Unlike for rough Co/Pt films, RPM is low at nucleation. However, as the magnetic field increases, RPM rapidly increases when the field approaches the remanent coercive point H_{cr} and plateaus at values as high as 80–90%. The plateau extends over a wide field region above the coercive point. RPM eventually decreases down to zero, when the field approaches saturation.

4.3. Cross-field MDM maps

In addition to measuring RPM, cross-field magnetic correlations were further carried out throughout the entire magnetization loop. Speckle patterns collected along the ascending branch and along the descending branch of the magnetization loop were cross-correlated. Resulting correlation maps $\rho(H_1, H_2)$ are shown in **Figure 13**, specifically the A × A map (correlations between two ascending branches) and the A × D map (correlations between ascending and descending branches) measured after one field cycle. The RPM and CPM information may be sliced from these map along their diagonal. Both the A × A and A × D correlation maps show high MDM, with ρ reaching as high as 95 ± 5% in the central region where $H_1 \approx H_2 \approx H_{cr}$. The high correlation forms a plateau extending throughout a wide region of field values from near above nucleation to near below saturation. This is not only true along the diagonal, where $H_1 = H_2$, but also off-diagonal where $H_1 \neq H_2$, revealing that the nanoscale magnetic domain patterns formed at these field values are all very similar [38].

4.4. Imprinting of a magnetic template via field cooling

The drastic behavioral differences between the MDM observed in ZFC [Co/Pd]/IrMn films and the MDM observed in rough Co/Pt films arise from the different microscopic magnetic interactions. In one case, MDM is induced by defects or disorder. In the other case, MDM is induced by exchange couplings. In this later case, MDM exits even in the absence of defects, that is, in smooth films. In the absence of defects, MDM is low at nucleation, as observed in **Figure 12**, because domains nucleate at random locations in the film. The reason for high MDM to occur at higher field values is the presence of a magnetic template imprinted in the AF layer during the cooling.

http://dx.doi.org/10.5772/intechopen.71076

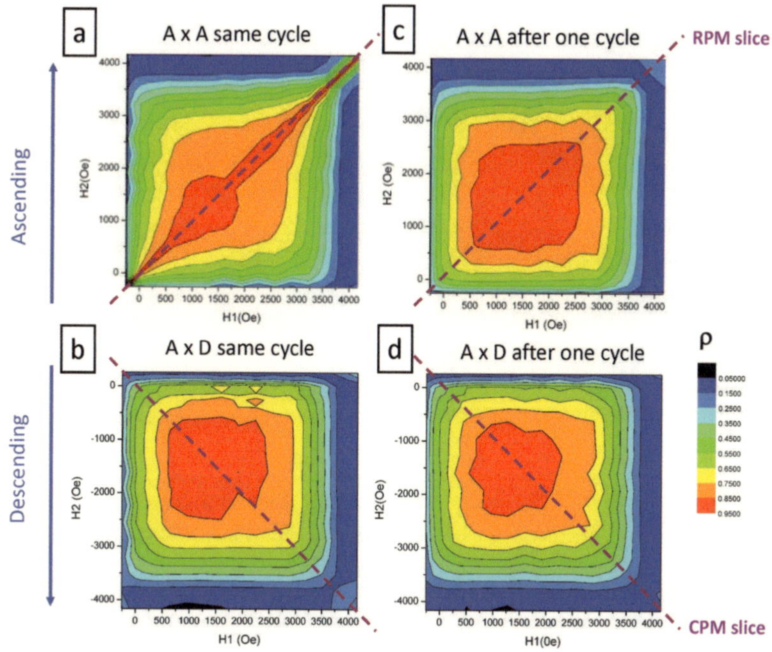

Figure 13. Cross-field MDM maps measured on [Co/Pd]/IrMn multilayers at 20 K in ZFC state. (a) A × A map measured on same cycle, (b) A × D map measured on same cycle, (c) A × A map measured after one cycle, (b) A × D map measured after one cycle. Extracted from Chesnel et al. [38].

In the specific material, the magnetic pattern gets imprinted from the F Co/Pd layer into the AF IrMn layer through exchange couplings via interfacial uncompensated spins. In ZFC state, the imprinted pattern is typically formed of interlaced magnetic domains with opposite magnetization, pointing perpendicular to the material, either out-of-plane or into the plane, as illustrated in **Figure 14a, b**. The area covered by one magnetization direction nearly equals the area covered by the other magnetization direction. When, in the ZFC state, the external magnetic field is cycled, the magnetic domains in the F layer successively nucleate, propagate, expand and eventually collapse at saturation. However, due to the frozen underlying magnetic pattern in the AF layer, the domain formation process in the F layer is not random (as it would be for a single smooth F layer). The domain formation process is highly guided by exchange coupling interactions with the underlying frozen template, so that when the remanent coercive point is reached, the domain pattern exactly matches the imprinted one [38, 44].

Guided by the exchange interactions with the magnetic template imprinted in the AF layer, the magnetic domain formation and reversal in the F layer is deterministic. The magnetic domains in the F layer always form in a way to match the underlying template. Consequently, high MDM occurs from about the remanent coercive point H_{cr}, where the template is fully matched, all the way up to nearly saturation, as domains evolve in a way to conserve the template topology. This happens both on the ascending and the descending branches of the magnetization loop, as illustrated in **Figure 14c**. Also the correlations between the ascending

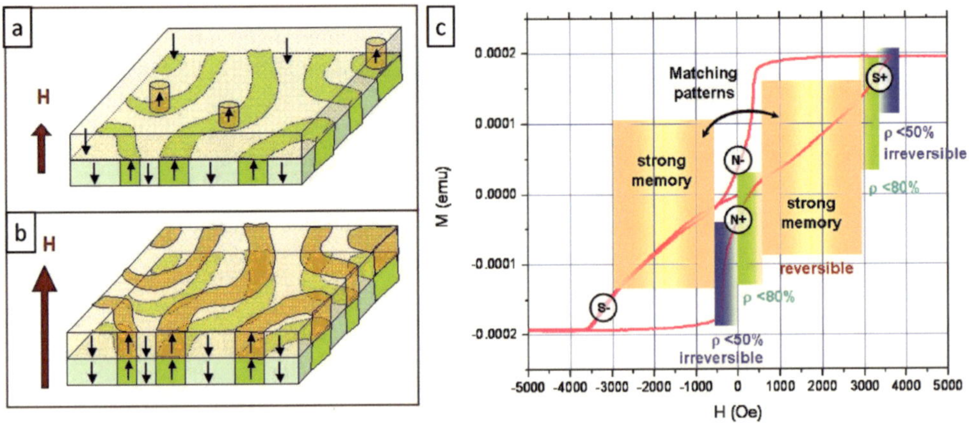

Figure 14. Illustration of the origin of MDM in exchange biased [Co/Pd]/IrMn multilayers. (a) and (b) Schematics of the domain pattern forming in the F layer (top) and AF layer (bottom) in a ZFC state where a striped domain pattern is imprinted in the AF layer (a) at nucleation, sparse bubble nucleate randomly in the F layer, (b) at the coercive point, the F domain pattern matches the AF imprinted pattern. (c) Schematics showing the occurrence of MDM throughout the magnetization cycle. MDM is the strongest in the central region of the magnetization loop from the remanent coercive point all the way up to near saturation, both on the ascending and descending branches of the magnetization loop. Extracted from Refs. [38, 44].

and descending branches is high. This is possible in the ZFC state because the imprinted pattern is made of long interlaced stripes with equal coverage of up and down domains.

5. Persistence of MDM through field cycling

When MDM is observed, a question that arises is if the memory persists through multiple field cycles. This question was investigated both in disorder-induced MDM and in exchange-bias induced MDM. In both cases, the magnetic correlations persisted through field cycling.

5.1. Field cycling dependence measurements

In the ergodic assumption, the measured correlation, which is an average over a statistical ensemble of microscopic magnetic domains, is expected to be the same than the time averaged magnetic correlation one would measure at any given location of the material. Because the area of the material probed by the x-rays typically includes thousands of magnetic domains and the system is at equilibrium, the ergodic assumption applies. It is therefore expected that the magnetic correlation measured between two magnetic states separated by a given number of cycles should not change in time.

For statistical purposes, correlation coefficients measured at the same number of separating loops but at different times have been averaged. For example, if speckle patterns were collected throughout five subsequent field cycles, the correlation coefficients between cycles

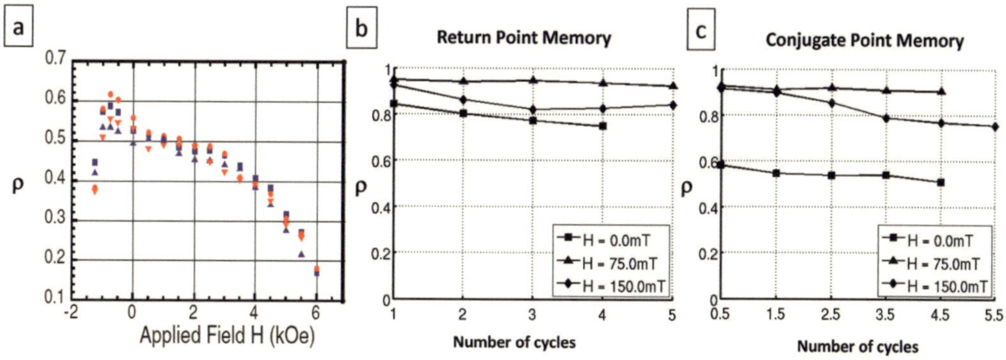

Figure 15. Persistence of MDM through field cycling. (a) RPM measured in Co/Pt multilayers after 1 cycle (red) and 11 cycles (blue). (b) and (c) MDM measured in exchange-biased [Co/Pd]/IrMn films (b) RPM vs. field cycle (c) CPM vs. field cycle. Extracted from Refs. [39, 45].

1 & 2, cycles 2 &3, cycles 3 & 4 and cycles 4 & 5 were all averaged, producing an average correlation coefficient corresponding to one separating cycle. Similar averages were applied for two separating cycles (averaging cycles 1 & 3, 2 & 4 and 3 & 5), three separating cycles, etc. Ultimately, the average correlation coefficient ρ was studied as a function of the number of separating cycles [45].

5.2. Dependence of MDM on number of field cycles

RPM and CPM correlations in [Co/Pd] multilayers and in [Co/Pd]/IrMn films were measured throughout many field cycles. In **Figure 15a**, the average RPM measured in Co/Pt films after 1 and after 11 cycles is plotted against the field H on the ascending branch of the magnetization loop. Even after 11 loops, RPM is as high as after one cycle [39]. In **Figure 15b**, **c**, the average RPM and CPM values measured in ZFC [Co/Pd]/IrMn films at specific points in field are plotted as a function of number of separating cycles. Both RPM and CPM appeared nearly constant as the number of separating cycles was increased. In particular, the optimal RPM value, which occurs near the coercive region, remains within 95–100%, and the optimal CPM in that region remains within 90–95%. The observed persistence of MDM with field cycling is consistent with predictions for exchange-bias induced magnetic memory. Indeed, in these systems, the magnetic domain template imprinted in the AF layer is frozen, meaning it does not change while the magnetic field is cycled, as long as the temperature is kept constant below the blocking point. The domain pattern in the F layer will therefore tend to always retrieve that same unchanged magnetic template, independently from the number of field cycles [45].

6. Spatial dependence of EC-induced MDM

The MDM results previously discussed in this chapter were obtained by cross-correlating entire speckle patterns altogether. The associated correlation numbers provided an ensemble average microscopic information, but no spatial dependency was probed. Spatial information is however included in the 2D speckle patterns which are being correlated for the estimation of MDM. This spatial information can be exploited to extract information about possible spatial dependency in MDM. Interesting oscillatory spatial dependence was found in the MDM exhibited by [Co/Pd]IrMn multilayers.

6.1. Exploring spatial dependence

Each 2D speckle pattern, such as the one in **Figure 7**, represents the intensity of the x-rays coherently scattered by the material. Because the scattering process involves an inversion from the real space to the scattering space, the spatial scale on the speckle images is an inverse of a distance. A common quantity to locate positions in the speckle patterns is the scattering vector q. The origin of the vector q is the center of the scattering pattern, as illustrated in **Figure 9**. The magnitude of q indicates spatial scales d being probed in the real space. q and d are linked by an inverse relationship: $q = \frac{2\pi}{d}$. The larger q is, the smaller the features in the real space. The spatial scales probed when collected scattering patterns such as the one in **Figure 7** typically range from about 50 nm up to a few microns.

Most scattering patterns observed when probing magnetic domain patterns in PMA films show a ring. The presence of the ring reveals a magnetic periodicity in the domain pattern with an isotropic arrangement (no preferred direction). The radius q^* of the ring represents the magnetic period d^*. This magnetic period is basically twice the domain width, as one period include a pair of up and down domains. In the [CoPd]/IrMn films, the observed magnetic period was typically $d^* \sim 300$–350 nm.

By cross-correlating selected regions of the scattering speckle pattern rather than the entire pattern, one obtains spatially dependent correlation numbers. This is useful to detect any spatial correlation features occurring at the nanoscale. For instance, magnetic domains patterns may correlate well at the short scale (100–500 nm), but not correlate so well in the long range (1 μm and above) or vice versa.

6.2. Space-field maps of MDM

To study the spatial behavior of MDM, in particular its dependence on q, cross-correlation is performed on selected rings of the speckle patterns. The rings are concentric, centered about the origin of the scattering pattern. The radius q of the ring is increased from zero to the largest size accessible in the image, thus providing a spatially dependent correlation coefficient $\rho(q)$ [40].

Because speckle patterns are collected at specific points in field H along the magnetization loop, the exploration of spatial dependence in MDM may be done at each field value H. Ultimately, the correlation ρ is mapped in a two-dimensional (q, H) space, thus probing the evolution of the spatial dependence in MDM with field. An example of such $\rho(q, H)$ is shown in **Figure 16a**.

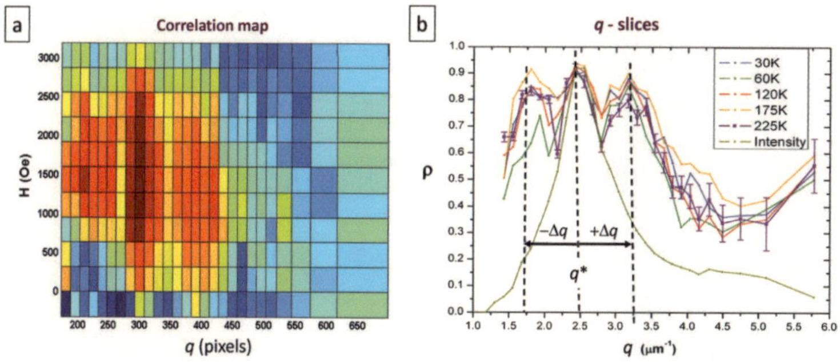

Figure 16. Q-dependence of MDM in ZFC exchange-biased [Co/Pd]/IrMn films. (a) $\rho(Q, H)$ correlation map measured at 30 K. (b) $\rho(q, H_c)$ slices through the correlation maps at $H = H_c$ measured at different temperatures from 30 K up to 225 K, showing an oscillation at all temperatures. Extracted from Chesnel et al. [46].

6.3. Oscillatory behavior of MDM in [Co/Pd] IrMn

The spatial and field dependence of MDM was explored in [Co/Pd] IrMn films [46]. The films were ZFC below the blocking temperature. Speckle patterns were collected at low temperature, throughout the magnetization cycle and several subsequent loops. Ring selective cross-correlations were carried out. The resulting $\rho(q, H)$ maps, averaged over subsequent cycles, showed interesting features. In particular, slices through the maps at specific H values around the remanent coercive point H_{cr} showed an oscillatory behavior for $\rho(q)$, as seen in **Figure 16b**.

The oscillation observed in the $\rho(q, H_c)$ curve in the ZFC [Co/Pd] IrMn films revealed spatially dependent MDM. The central peak in the $\rho(q, H_c)$ curve seen on **Figure 16b**, occurred at the same location q^* than the ring in the scattering pattern. This suggested that MDM is strongly correlated with the magnetic period in the magnetic domain pattern, here around 400 nm. Two satellite peaks were observed around the central peak, at the same distance ± Δq from q^*. The presence of these satellite peaks suggested a spatial superstructure in MDM. The size of the superstructure, set by the value Δq, was found to be about $D \sim 1.5$ μm. Topographical AFM images of the surface of the films indicated that D nearly matched the average distance between structural defects in the material.

6.4. Azimuthal angular dependence

Since the speckle patterns used for the evaluation of MDM are 2D images, one can explore the spatial dependence of MDM in at least two directions. In addition to probing the dependence on the scattering vector q, one can investigate the angular dependence, while varying the azimuthal angle Δ that indicates a particular location on the ring [47], as illustrated in **Figure 17**. In that case, cross-correlations are measured between two points on the ring separated by an angle Δ, leading to a correlation coefficient $\rho(q, \Delta)$. Such study was carried out on ZFC [Co/Pd] IrMn films and showed some periodical variations as a function of the angle Δ as illustrated in **Figure 17c**. These observations suggested the existence of some hidden rotational symmetries in the formation of the disordered magnetic domain patterns in these films.

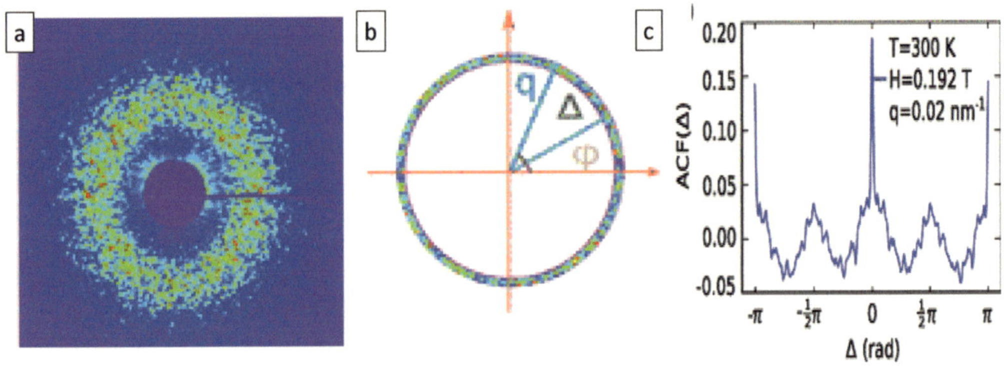

Figure 17. Angular dependence of MDM in ZFC exchange-biased [Co/Pd]/IrMn films. (a) Speckle pattern collected near the coercive point. (b) Selected ring from the speckle pattern (a). The angles θ and Δ are defined here. (c) Autocorrelation function plotted against Δ. Extracted from Su et al. [47].

7. Dependence of EC-induced MDM on temperature

The results discussed in the previous sections show the occurrence of MDM via exchange bias when the film is cooled down to low temperatures, well below the blocking temperature T_B. Most of the measurements were performed at around 20 K after ZFC cooling the material. A question that arises next is if the EC-induced MDM depends on temperature and how it behaves at the phase transition near T_B. These questions were investigated in exchange biased [Co/Pd]IrMn films [38].

7.1. Probing temperature dependence in zero-field-cooled (ZFC) state

To probe the temperature dependence of MDM, the [Co/Pd]/IrMn films were first heated up to around 400 K, well above the blocking temperature $T_B \sim 300$ K and demagnetized at 400 K. The films were then ZFC down to 20 K. Speckle patterns throughout many magnetization cycles were collected in the ZFC state at 20 K, and then at higher temperatures while warming the film all the way back up to above T_B. Average A × A and A × D correlation maps were measured at several points in temperature. For each point, the temperature was finely controlled and stabilized via a cryogenic environment. The resulting correlation maps are shown in **Figure 18**.

7.2. Persistence of MDM through warming up

The correlation maps in **Figure 18a** show the occurrence of high MDM extending throughout a wide region of the magnetization loop at all temperatures below T_B. The high MDM occurred for both A × A and A × D correlation maps between same branches or opposite branches of the magnetization loop, respectively. At all temperatures below T_B, MDM reached a high value plateauing over a large region of field values around the coercive point.

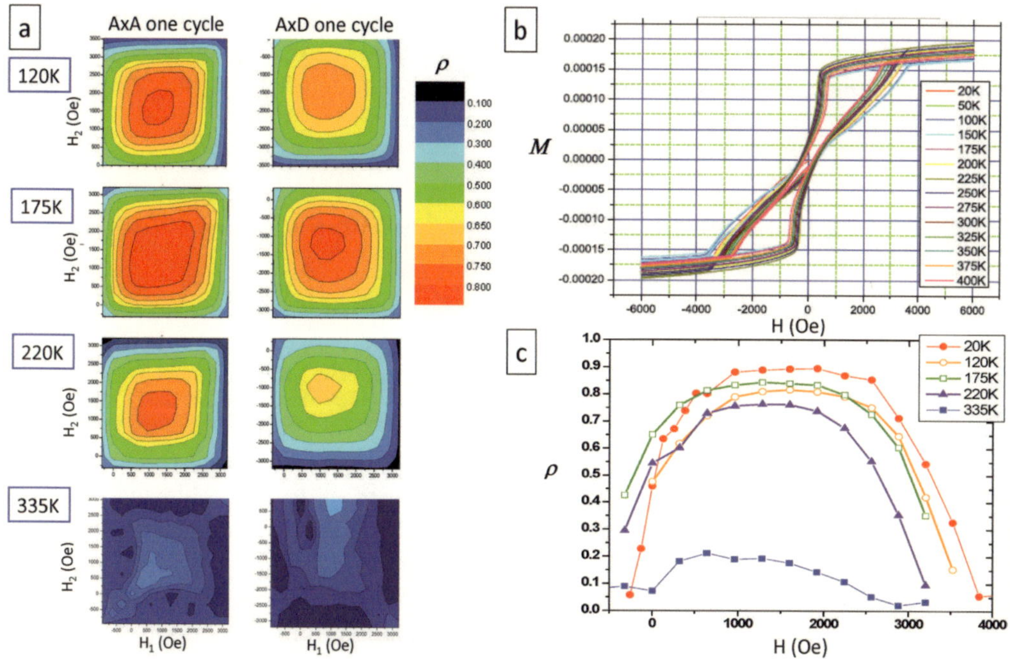

Figure 18. Temperature dependence of MDM in ZFC exchange-biased [Co/Pd]/IrMn films. (a) A × A and A × D correlation maps measured after one cycle at different temperatures while warming from 20 K up to 335 K. (b) Magnetization loop at various temperatures. (c) Slices through the A × A maps at $H \approx H_{cr}$. Extracted from Chesnel et al. [38].

The temperature dependence of MDM may be compared to the temperature dependence of the magnetization loop. The shape of the magnetization loop in the ZFC state, as seen in **Figure 18b** is symmetrical, centered about the origin. The loop has an hourglass-like shape; it is narrow at the center and opens up at the extremities, due to the presence of exchange couplings. When increasing the temperature from the ZFC state, the overall shape of the magnetization loop remains the same, but the magnitude of the opening, or hysteresis, progressively decreases. Despite the changes observed in the magnetization loop while warming the film up from the ZFC state, MDM remains strong and extended at all temperatures below T_B. Slices through the correlation maps, shown in **Figure 18c**, all show the same trend: low correlation at nucleation, sharply increasing to reach a high correlation plateau in the central region of the magnetization loop, and then sharply decreasing toward saturation.

7.3. Disappearance of MDM above the blocking temperature

When the temperature is increased above T_B, MDM vanishes, as shown in **Figure 18**. At that stage, the correlation map shows nearly zero correlation, becoming all blue.

The persistence of high MDM throughout warming, at all temperatures below T_B, and its vanishing above T_B confirms that MDM is here purely induced by exchange-couplings between the F and AF layers. Above T_B, these exchange couplings disappear. Consequently, the magnetic domain template which was imprinted in the AF layer throughout the uncompensated spins at the interface is lost once the magnetic field is cycled again. In the absence of magnetic template, and since the film is smooth (absence of defects), the magnetic domains in the F layer nucleate and propagate randomly.

8. Optimizing EC-induced MDM by adjusting field cooling conditions

All the studies of EC-induced MDM discussed in the previous sections were carried out on zero-field-cooled [Co/Pd]/IrMn films. Next question is what happens if the film is now field-cooled (FC), i.e., cooled in the presence of a magnetic film. Will MDM remain as high in the FC state as it is in the ZFC state?

This question was investigated by studying the magnetic correlations after cooling the [Co/Pd]/IrMn films under various cooling field conditions. The resulting correlation maps showed drastic changes as a function of the magnitude of the cooling field [33]. The result of this study is summarized in **Figure 19**. Magnetic correlations were measured in [Co/Pd]/IrMn films after field cooling the material under magnetic field of various magnitudes H_{FC}, increasing from zero (ZFC state) all the way up to high values, near saturation value H_s.

8.1. Dependence of MDM on cooling field magnitude

When the magnitude H_{FC} of the cooling field is near zero, high MDM, up to 100%, is observed. This high MDM extends from nucleation (lower left corner of the correlation map) all the way up to saturation (upper right corner of the correlation map).

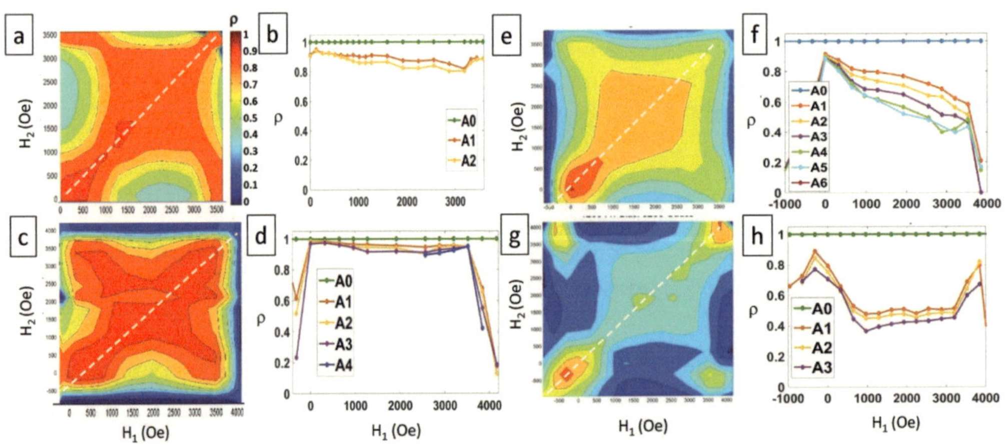

Figure 19. Cooling condition dependence of MDM in [Co/Pd]/IrMn films. (a) A × A correlation map measured at 20 K after ZFC; (b) slices through map (a); (c) A × A map measured at 20 K after cooling under H_{FC} = 2240 Oe; (d) slices through map (c); (e) A × A map measured at 20 K after cooling under H_{FC} = 2560 Oe; (f) slices through map (e); (g) A × A map measured at 20 K after cooling under H_{FC} = 3200 Oe; (h) slices through map (g). Extracted from Chesnel et al. [33].

When H_{FC} is increased up to moderately high values, high MDM is still observed, with a wide correlation plateau extending throughout a large range of field values from nucleation to saturation.

When H_{FC} approaches the saturation value, which for the [Co/Pd]/IrMn film is around $H_s \approx 3200$ Oe, the measured magnetic correlations rapidly decrease. At $H_{FC} \approx 2600$ Oe, the correlation drops from about 90% at nucleation down to about 50% in the central region of the magnetization loop.

When $H_{FC} \approx H_s \approx 3200$ Oe, MDM has almost vanished, except at the nucleation and saturation extremities of the magnetization loop.

8.2. Shaping MDM by adjusting cooling field conditions

In the ZFC state, the magnetic domain template imprinted in the AF layer is formed of long interlaced stripes with about the same amount of domains of opposite magnetizations (about a 50% up: 50% down split). The imprinted domain pattern constitutes a template for the domain reversal in the F layer. Because of exchange couplings occurring between the Co spins in the F layer and the uncompensated interfacial Mn spins in the AF layer, the domains form in a way to match the underlying template. The matching occurs rapidly when approaching the coercive point (where the up-down magnetization split is exactly 50–50%). The matching persists at higher field values, all the way up to near saturation. This is possible because, even though the magnetic domains of opposite magnetization expand and shrink, the specific topology of the domain pattern still matches the underlying template, as illustrated in **Figure 20**.

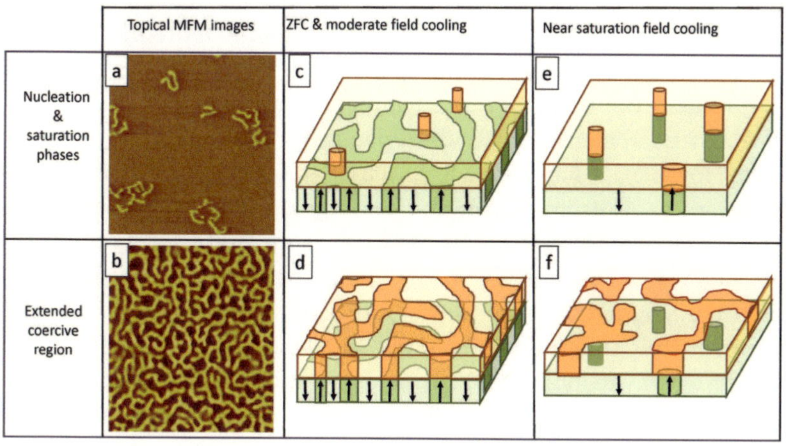

Figure 20. Illustration of the magnetic domain memory effect in the exchange-biased [Co/Pd]/IrMn film under various field cooling conditions. The imprinted pattern in the AF IrMn layer is shown in green color, the domain pattern in the F layer is shown in orange color. Arrows indicate magnetization direction. (a, b) 5 × 5 μm MFM images showing actual magnetic domains in the F Co/Pd layer at room temperature: (a) near nucleation and (b) near the coercive point. (c, d) Sketches of the magnetic domain configuration for ZFC and moderate field-cooled states: (c) near nucleation and (d) near the coercive point. (e, f) Sketches of the magnetic domain configuration in near-saturating field-cooled states: (e) near nucleation and (f) near the coercive point. Extracted from Chesnel et al. [33].

In the FC states, the magnetic domain template imprinted in the film has an unbalance of up and down domains, and the net magnetization is non-zero. However, if the magnitude H_{FC} of the cooling field remains in a certain range above H_{cr}, the imprinted pattern, still formed of interlaced up and down domains, provides a template that entirely drives the reversal of the magnetic domains in the F layer. The topology of the imprinted pattern in the FC state resembles that of the imprinted pattern in the ZFC state. High MDM is therefore maintained throughout the entire magnetization process.

If the magnitude H_{FC} of the cooling field approaches saturation value H_s, the imprinted magnetic domain pattern does not include long interlaced magnetic domains of opposite magnetization anymore but a few bubble domains sparsely scattered throughout the film. This imprinted template is not able to drive the magnetic domain formation throughout the entire reversal but only at the extremities, that are the nucleation and the saturation points, as illustrated in **Figure 20**.

These results demonstrate the possibility to induce and control nanoscale MDM in exchange biased films by adjusting the field cooling conditions. High, up to 100% MDM, extending throughout the entirety of the magnetization loop can be achieved by cooled under no field or relatively low field values. However, MDM can be almost eliminated, by cooling the material under high magnetic field, approaching saturation and higher.

9. Conclusion

Magnetic domain memory (MDM) is an unusual property exhibited by certain ferromagnetic films, where the microscopic magnetic domains tend to reproduce the same topological pattern after it has been erased by an external magnetic field. In most ferromagnetic materials, MDM does not occur. When an external magnetic field is applied and cycled, microscopic magnetic domains form and propagate throughout the material in non-deterministic ways. However, it has been found that some ferromagnetic thin films with perpendicular magnetic anisotropy (PMA) do show significant MDM under certain structural and magnetic conditions. One structural condition is the presence of defects. When rough enough, thin Co/Pt multilayers with PMA, exhibits partial MDM occurring in the nucleation phase of the magnetization process. This disorder-induced MDM is caused by the presence of microscopic structural defects, playing the role of pinning sites for the domain nucleation. Another way to induce MDM and to maximize it, even in smooth films, is incorporating magnetic exchange couplings (EC) between a PMA ferromagnetic (F) film and an underlying antiferromagnetic (AF) film. This has been achieved by combining F [Co/Pd] multilayers with AF IrMn layers. After demagnetizing the material and zero-field-cooling it below its blocking temperature, high MDM up to 100% was observed throughout almost the entirety of the magnetization process. This high MDM is induced by EC interactions between the Co spins in the F layer and the interfacial uncompensated spin in the IrMn layer. When the material is cooled down below its blocking temperature, a specific magnetic domain pattern gets imprinted into the AF layer. When the field is cycled at low temperature, the frozen imprinted AF pattern then plays the role of a template for the domain formation in the F layer. The resulting high MDM extends throughout a wide range of field values, from the coercive point to nearly saturation. This EC-induced MDM persists through field cycling and through warming the material all the way up the blocking temperature, above which MDM vanishes. Additionally, it was found that the amount of EC-induced MDM can be varied by adjusting the magnitude of the field applied during the cooling. If the material is cooled under no field or moderate field, MDM reaches high values up to 100% throughout most of the magnetization process. If, however, the material is cooled under a nearly saturating field, MDM vanishes, except at the nucleation and saturation extremities of the magnetization cycle.

These observations of MDM in certain PMA ferromagnetic films opens the door to more investigations. A particular question is if there are other ways to induce and control MDM in a material. It has been recently found that EC-induced MDM may be affected by light, such as x-rays. If the material is illuminated by too intense x-rays, the film may lose its EC properties and MDM may vanish. This finding, which resonates with the emergent all-optical magnetic switching phenomena observations [48–50], suggest that EC-induced MDM could be controlled by light illumination. Ultimately, the ability to induce and control MDM in PMA ferromagnetic film, either by structural disorder, or by exchange couplings and light illumination, may offer a tremendous potential for improving technological applications in the field of magnetic recording and spintronics.

Author details

Karine Chesnel

Address all correspondence to: kchesnel@byu.edu

Department of Physics & Astronomy, Brigham Young University (BYU), Provo, UT, USA

References

[1] Néel L. Les lois de l'aimantation et de la subdivision en domaines élémentaires d'un monocristal de fer. Journal de Physique et Le Radium. 1944;**5**:241

[2] Kittel C. Theory of the Structure of Ferromagnetic Domains in Films and Small Particles. Physics Review. 1946;**70**:965

[3] Kittel C. Physical theory of ferromagnetic domains. Reviews of Modern Physics. 1949;**21**:541

[4] Fowler Jr CA. Fryer EM. Magnetic domains of thin films of Nickel-Iron. Physics Review. 1955;**100**:746

[5] Williams HJ, Sherwood RC. Magnetic domains in thin films. Journal of Applied Physics. 1957;**28**:548

[6] Chapman JN. The investigation of magnetic domain structures in thin foils by electron microscopy. Journal of Physics D: Applied Physics. 1984;**17**:623

[7] Allenspach R, Stampanoni M, Bischof A. Magnetic domains in thin epitaxial Co/Au(111) films. Physical Review Letters. 1990;**65**:3344

[8] Lecoeur P, et al. Magnetic domain structures of LaSrMnO3 thin films with different morphologies. Journal of Applied Physics. 1997;**82**:3934

[9] Vedmedenko EY, Ghazali A, Levy J-CS. Magnetic vortices in ultrathin films. Physical Review B. 1999;**59**:3329

[10] Wachowiak A, et al. Direct observation of internal spin structure of magnetic vortex cores. Science. 2002;**298**:577

[11] Shigeto K, Okuno T, Mibu K, Ono TS. Magnetic force microscopy observation of antivortex core with perpendicular magnetization in patterned thin film or permalloy. Applied Physics Letters. 2002;**80**:4190

[12] Yu XZ, et al. Near room- temperature formation of a skyrmion crystal in thin-films of the helimagnet FeGe. Nature Materials. 2011;**10**:106

[13] Wilson MN, et al. Extended elliptic skyrmion gratings in epitaxial MnSi thin films. Physical Review B. 2012;**86**:144420

[14] Fert A, Cros V, Sampaio J. Skyrmions on the track. Nature Nanotechnology. 2013;**8**:152

[15] Jiang W, et al. Blowing magnetic skyrmion bubbles. Science. 2015;**349**:283

[16] Johnson MT, Bloemen PJH, den Broeder FJA, de Vriest JJ. Magnetic anisotropy in metallic multilayers. Reports on Progress in Physics. 1996;**59:**1409

[17] Gehanno V, Samson Y, Marty A, Gilles B, Chamberod A. Magnetic susceptibility and magnetic domain configuration as a function of the layer thickness in epitaxial FePd (001) thin films ordered in the L10 structure. Journal of Magnetism and Magnetic Materials. 1997;**172**:26

[18] Gehanno V, Marty A, Gilles B, Samson Y. Magnetic domains in epitaxial ordered FePd (001) thin films with perpendicular magnetic anisotropy. Physical Review B. 1997;**55**:12554

[19] Thiele J-U, Folks L, Toney MF, Weller DK. Perpendicular magnetic anisotropy and magnetic domain structure in sputtered epitaxial FePt (001) L10 films. Journal of Applied Physics. 1998;**84**:5686

[20] Hellwig O, Denbeaux GP, Kortright JB, Fullerton EE. X-ray studies of aligned magnetic stripe domains in perpendicular multilayers. Physica B. 2003;**336**:136

[21] Hellwig O, Berger A, Kortright JB, Fullerton EE. Domain structure and magnetization reversal of antiferromagnetically coupled perpendicular anisotropy films. Journal of Magnetism and Magnetic Materials. 2007;**319**:13

[22] Richter HJ, Harkness IV SD. Media for magnetic recording beyond 100 Gbit/in.2. MRS Bulletin. 2006;**31**:384

[23] Chappert C, Renard D, Beauvillain P, Renard JP, Seiden J. Ferromagnetism in very thin films of Nickel and Cobalt. Journal of Magnetism and Magnetic Materials. 1986;**54**:795

[24] Carcia PF. Perpendicular magnetic anisotropy in Pd/Co and Pt/Co thin-film layered structures. Journal of Applied Physics. 1988;**63**:5066

[25] Lin CJ, Gorman GL, Lee CH, Farrow RFC, Marinero EE, do HV, Notarys H, Chien CJ. Magnetic and structural properties of Co/Pt multilayers. Journal of Magnetism and Magnetic Materials. 1991;**93**:194

[26] Winkler G, Kobs A, Chuvilin A, Lott D, Schreyer A, Oepen HP. On the variation of magnetic anisotropy in Co/Pt(111) on silicon oxide. Journal of Applied Physics. 2015;**117**:105306

[27] Chesnel K, Westover AS, Cardon KH, Dodson B, Healey M, et al. Topological bubble stripe phase transition in remanent magnetic domain patterns in Co/Pt thin films and its dependence with Co thickness. to appear

[28] Westover AS, Chesnel K, Hatch K, Salter P, Hellwig O. Enhancement of magnetic domain topologies in Co/Pt thin films by fine tuning the magnetic field path throughout the hysteresis loop. Journal of Magnetism and Magnetic Materials. 2016;**399**:164

[29] Kappenberger P, Martin S, Pellmont Y, Hug HJ, Kortright JB, Hellwig O, Fullerton EE. Direct imaging and determination of the uncompensated spin density in exchange-biased CoO/ (CoPt) multilayers. Physical Review Letters. 2003;**91**:267202

[30] Chesnel K, Belakhovsky M, Livet F, Collins SP, van der Laan G, Dhesi SS, Attané JP, Marty A. Soft x-ray magnetic speckles from a nanostructured FePd wire. Physical Review B. 2002;**66**:172404

[31] Chesnel K, Belakhovsky M, van der Laan G, Livet F, Marty A, Beutier G, Collins SP, Haznar A. Tracking the local reversal processes in nanostructures by magnetic speckles. Physical Review B. 2004;**70**:180402(R)

[32] Eisebitt S, Luning J, Schlotter WF, Lorgen M, Hellwig O, Eberhardt W, Stohr J. Lensless imaging of magnetic nanostructures by x-ray spectro-holography. Nature. 2004;**432**:885

[33] Chesnel K, Safsten A, Rytting M, Fullerton EE. Shaping nanoscale magnetic domain memory in exchange-coupling ferromagnets by field cooling. Nature Communications. 2016;**7**:11648

[34] Chesnel K, Turner JJ, Pfeifer M, Kevan SD. Probing complex materials with coherent soft x-rays. Applied Physics A: Materials Science & Processing. 2008;**92**:431

[35] de Bergevin F, Brunel M. Diffraction of x-rays by magnetic materials. I. General formulae and measurements on ferro- and ferrimagnetic compounds. Acta Cryst. 1981;**A37**:314

[36] Hannon JP, Trammell GT, Blume M. D, Gibbs. X-ray resonance exchange scattering. Physical Review Letters. 1988;**61**:1245

[37] Kortright JB, Awschalom DD, Stohr J, Bader SD, Idzerda YU, Parkin SSP, Schuller I, Siegmann H-C. Research frontiers in magnetic materials at soft x-ray synchrotron radiation facilities. Journal of Magnetism and Magnetic Materials. 1999;**207**:7

[38] Chesnel K, Wilcken B, Rytting M, Kevan SD, Fullerton EE. Field mapping and temperature dependence of magnetic domain memory induced by exchange couplings. New Journal of Physics. 2013;**15**:023016

[39] Pierce MS, Moore RG, Sorensen LB, Kevan SD, Hellwig O, Fullerton EE, Kortright JB. Quasistatic x-ray speckle metrology of microscopic magnetic return-point memory. Physical Review Letters. 2003;**90**:175502

[40] Chesnel K, Nelson J, Wilcken B, Kevan SD. Mapping spatial and field dependence of magnetic domain memory by soft x-ray speckle metrology. Journal of Synchrotron Radiation. 2012;**19**:293

[41] Pierce MS, Buechler CR, Sorensen LB, Turner JJ, Kevan SD, Jagla EA, Deutsch JM, Mai T, Narayan O, Davies JE, liu K, Dunn JH, Chesnel KM, Kortright JB, Hellwig O, Fullerton EE. Disorder-induced microscopic magnetic memory. Physical Review Letters. 2005;**94**:017202

[42] Pierce MS, Buechler CR, Sorensen LB, Kevan SD, Jagla EA, Deutsch JM, Mai T, Narayan O, Davies JE, Liu K, Zimanyi GT, Katzgraber HG, Hellwig O, Fullerton EE, Fischer P, Kortright JB. Disorder-induced magnetic memory: experiments and theories. Physical Review B. 2007;**75**:144406

[43] Seu KA, Su R, Roy S, Parks D, Shipton E, Fullerton EE, Kevan SD. Microscopic return point memory in Co/Pd multilayer films. New Journal of Physics. 2010;**12**:035009

[44] Chesnel K, Fullerton EE, Carey MJ, Kortright JB, Kevan SD. Magnetic memory in ferromagnetic thin films via exchange coupling. Physical Review B. 2008;**78**:132409

[45] Nelson J, Wilcken B, Chesnel K. Persistence of magnetic domain memory through field cycling in exchange bias thin films. Journal of the Utah Academy of Sciences, Arts & Letters. 2010;**87**:267

[46] Chesnel K, Nelson JA, Kevan SD, Carey MJ, Fullerton EE. Oscillating spatial dependence of domain memory in ferromagnetic films mapped via x-ray speckle correlation. Physical Review B. 2011;**83**:054436

[47] Su R, Seu KA, Parks D, Kan JJ, Fullerton EE, Roy S, Kevan SD. Emergent rotational symmetries in disordered magnetic domain patterns. Physical Review Letters. 2011;**107**:257204

[48] Stanciu CD, Hansteen F, Kimel AV, Kirilyuk A, Tsukamoto A, Itoh A, Rasing T. All-optical magnetic recording with circularly polarized light. Physical Review Letters. 2007;**99**:047601

[49] Mangin S, Gottwald M, C-H Lambert, Steil D, Uhlir V, Pang L, Hehn M, Alebrand S, Cinchetti M, Malinowski G, Fainman Y, Aeschlimann M, Fullerton EE. Engineered materials for all-optical helicity-dependent magnetic switching. Nature Materials. 2014;**13**:286

[50] Shcherbakov M, Vabishchevich PP, Shorokhov AS, Chong KE, Choi D-Y, Staude I, Miroshnichenko AE, Neshev DN, Fedyanin AA, Kivshar YS. Ultrafast all-optical switching with magnetic resonances in nonlinear dielectric nanostructures. Nano Letters. 2015;**15**:6985

Section 3

Magnetism in Alloys

Chapter 4

Magnetic Characteristics of High Entropy Alloys

Rajesh K. Mishra and Rohit R. Shahi

Additional information is available at the end of the chapter

http://dx.doi.org/10.5772/intechopen.74235

Abstract

High entropy alloy (HEA) is a multi-principal alloy having at least five principal elements in the concentration range of 5–35 at.%. HEAs having excellent mechanical properties and further these properties can be altered by the addition of different alloying element. For example with the addition of Al in base alloy make them a ductile and the addition of Co, Ti, etc. transforms base alloy to brittle material. This characteristic of HEAs makes them a promising technologically important material. A soft magnetic material should have good mechanical property, structural stability at high temperature and low coercivity with high magnetization. Recently, reported $FeCoNiMn_{0.25}Al_{0.25}$ and CoCrFeNiM (M = Cu, Mn) HEAs got attention as a better soft magnetic material because these HEAs having good soft magnetic characteristics along with good mechanical and excellent structural stability at high-temperature. Recent reports described that the mechanical as well as magnetic characteristics of these alloys can be tuned by the variation and/or the addition of alloying element in the base alloys. The magnetic characteristics of these alloys basically depend on the alloying element and compositional variation of the magnetic element present in particular HEAs. We have summarized the key results of magnetic characteristics of some recently investigated promising high entropy alloys.

Keywords: high entropy alloy, magnetic properties, soft magnetic materials, alloying

1. Introduction

For centuries alloy design has been a hot topic for metallurgists and material scientists. The design concepts of alloys is based on only one or two principle elements, while minor fraction of other elements are added for property enhancement. This classical definition has been broken by the Yeh et al. in 2004 by the suggestion of a new alloy design concept, referred as multicomponent alloys named as high entropy alloys (HEAs) [1]. They defined the HEAs in two ways one is based on composition and other on entropy. Under the compositional definition, HEAs are defined as those alloys having at least 5 principle elements [4]. The each with a concentration range of 5–35 at.% [1–4]. From the entropic point of view, the HEAs are defined as those alloys having configurational entropy larger than 1.5R (where R is the gas constant) [1–4]. Moreover, it should be noted that rather than the principle element in HEAs, some minor element (having concentration less than 5 at.%) have also been added to optimize the properties presence of multiple principal elements in HEAs is not merely a compositional difference compared to the conventional alloys, but it has a fundamental effect on the system characteristics that leads four core effects: high entropy, severe lattice distortion, sluggish diffusion and cocktail effects [2–4]. Due to these characteristics it has been reported that HEAs are a good candidate of structural as well as functional engineering materials that have high strength and hardness, excellent corrosion and wear resistance, as well as great fatigue resistance [2–6].

Previous studies discovered that HEAs generally exhibit simple solid solution structures such as body centered cubic (BCC), face centered cubic (FCC), hexagonal close packed (HCP) or mixture of the above rather than the intermetallic compounds due to the high entropy effect. However, minor fractions of intermetallic phases are observed in some HEAs [7]. The formation of solid solution phases in HEAs are governed by several thermodynamic parameters. These are entropy of mixing (ΔS_{mix}), enthalpy of mixing (ΔH_{mix}) and atomic size difference (δ) [8]. These parameters are defined as follows:

$$\Delta S_{mix} = -R\sum_{i=1}^{n} Ci \ln Ci \tag{1}$$

Here, R is the gas constant and C_i is the molar fraction of ith element.

$$\Delta H_{mix} = \sum_{i=1}^{n} 4\Delta H_{ij}^{mix} CiCj \tag{2}$$

Here, ΔH_{ij}^{mix} is the enthalpy of mixing of the binary liquid between the ith and j^{th} element and C_i, C_j are the atomic fraction of ith and j^{th} elements.

$$\delta\% = 100\%\sqrt{\sum_{i=1}^{n} Ci\left(1-\frac{r_i}{\sum_{j=1}^{n} C_j r_j}\right)^2} \tag{3}$$

Here, C_i and r_i denotes the atomic fraction and atomic radius of the ith element respectively.

Guo et al. reported that HEAs forms simple solid solution phases when $-22 \leq \Delta H_{mix} \leq 7$ kJ/ mol, $11 \leq \Delta S_{mix} \leq 19.5$ J/K mol and $\delta \leq 8.5\%$. Moreover, it has been found that the most critical factor that decides whether the formed simple solid solution phases are FCC or BCC in HEAs is the valence electron concentration (VEC) [9]. It has been observed that for VEC < 6.87 and VEC ≥ 8.0 BCC and FCC phases are formed respectively, while both phases (i.e., mixture of FCC and BCC) will coexist when $6.87 \leq \text{VEC} \leq 8.0$ [9].

The typical synthesis routs used for traditional alloys can also be applied for the synthesis of HEAs. On the dependence of how the constituent elements are mixed, the processing routs are classified into four categories [10–13]: (1) from the liquid state (arc melting), (2) from solid state (mechanical alloying), (3) from the gas state (sputtering), and (4) from electrodeposition process. It has been found that for the same composition of the alloys, different synthesis rout produced distinct microstructure that leads different properties. For example, CoCrFeNiTi HEA synthesized through arc melting has mixture of FCC, BCC and minor fraction of an intermetallic phase [14]. However, when it was synthesized through mechanical alloying (MA), it has only single FCC phase [15]. In addition to the synthesis rout, selection and composition of constituent elements also plays a critical role in deciding the microstructure and phase composition that influenced the properties of the HEAs [16–18]. Addition of Ti into CoCrFeNi leads the formation of intermetallic phases such as Laves and σ-phases and enhanced the hardness and compressive strength [16]. Moreover, Shun et al. reported that as the content of Mo increased from 0 to 0.85 into $CoCrFeNiMo_x$, the crystal structure changed from single FCC to mixture of FCC with (Mo, Cr)-rich σ and μ phases, correspondingly the hardness and compressive strength of the system increased up to 68 and 40% as compared to initial sample [17]. Liu et al. also reported that addition of Nb into CoCrFeNi changes the microstructure from the initial single FCC to mixture of FCC+ Nb-rich Laves phase. It has been found that the fracture and yield strength of the system increased from 413 to 1004 MPa and 147 to 637 MPa with the addition of Nb [18]. Apart from the mechanical property of the HEAs, functional properties of HEAs such as magnetic property, electrical resistance, etc. attracted great attention during last 5 years [19]. Recent reports showed that the soft magnetic characteristics along with optimal mechanical behaviour of HEAs can be obtained through carefully selection of constituent elements [20–22]. The magnetic characteristics of the alloys are sensitive to the base alloy, addition of alloying elements, resulting phases and their volume fractions [23, 24]. For example, Zuo et al. reported that the addition of Si in CoFeNi reduced the value of Ms more significantly from 151.3 to 80.5 emu/g than Al (from 151.3 emu/g to 101.8 emu/g) [23]. Moreover, Kao et al. reported that the magnetic nature of $Al_xCoCrFeNi$ HEA turned from paramagnetic to ferromagnetic, when the crystal structure changed from FCC (at $x = 0$) to BCC (at $x = 2.0$) [24]. Thus as a consequence the selection of alloying elements, their composition and the type of resultant crystal structure are the dominant factor for controlling the mechanical as well as magnetic properties of HEAs. In this chapter we have discussed and described the magnetic behaviour of recently developed multicomponent HEAs.

2. Effects of constituent elements on magnetic characteristics of different HEAs

Soft magnetic materials are widely used in the electrical power generation and transmission, electric motors, electromagnets, etc. However, conventional alloys have certain limitations, such as brittleness, weak mechanical behaviour and low electrical resistivity of Fe—Ni alloys. However, some FeCoNi based HEAs have better soft magnetic characteristics along with excellent mechanical properties [20–22, 25]. These combined properties of FeCoNi based HEAs are rarely seen in conventional alloys. Moreover, it has been found that the optimal balance of magnetic and mechanical properties of HEAs is dominantly affected by the alloying elements and their composition [20–22, 25]. Among all the studied alloys the ternary equiatomic FeCoNi alloy having single FCC structure exhibits higher saturation magnetization (Ms) (151 emu/g) with low coercivity (Hc) (1.52 Oe) [23]. Any addition of other paramagnetic or diamagnetic elements will affect its resultant magnetic characteristics. Since, it is well known that the magnetic characteristics of the alloys are very sensitive to the composition, alloying elements and the resultant phase structure [26]. For example, addition of Cr in FeCoNi base alloy changed its magnetic nature from ferromagnetic to paramagnetic. However, further addition of Al or Pd in CoCrFeNi alloy, the paramagnetic nature of base alloy changed to ferromagnetic. It has been reported that FeCoCrNiAl alloy has the value of the Ms = 13 emu/g and value of Ms for $FeCoCrNiPd_2$ alloy was found to be 34 emu/g [26]. In addition to the increase in saturation magnetization, Lucas et al. also found that with the addition of Al and Pd in CoCrFeNi alloy, the Curie temperature increased up to 157 and 230°C respectively. Enhancement of Curie temperature with Pd addition makes this alloy useful for magnetic refrigeration near room temperature [26].

Zhang et al. investigated the magnetic and mechanical properties of most widely studied equiatomic AlCoCrCuFeNi HEA in as cast and annealed state [27]. They reported that both alloys exhibit soft magnetic characteristics. Moreover, after annealing, the value of Ms and Hc decreased from 38.18 to 16.08 emu/g and 45 to 15 Oe respectively. The decrement of Ms is associated to the structure coarsening and phase transformation [27]. For detail study, Singh et al. investigate the microstructure of AlCoCrCuFeNi HEA in three different states (splat-quenched, as cast and aged at 600°C) [28]. They reported that the ferromagnetic characteristic of HEAs is correlated to the decomposition of the Cr—Fe—Co rich regions into Fe—Co-rich and Cr-rich regions. It has been found that splat-quenched alloy exhibit better soft magnetic characteristic than the as cast and aged HEAs. This was associated to the initial stage decomposition of Cr—Fe—Co region, resulting in clustering of Fe-rich and Cr-rich domain. However, the aged alloy exhibit high value of Ms, Hc and remanence ratio as compared to the as cast HEA due to a higher degree of decomposition in Cr—Fe—Co rich regions. Additionally, they also pointed out that the aged alloy exhibits no ferromagnetic phase transition up to temperature of 400 K at 1 T [28].

Wang et al. investigated the effect of milling duration and composition on microstructure and magnetic properties of FeSiBAlNi and FeSiBAlNiNb HEAs [29]. Initially amorphous HEA has been synthesized by MA method. They found that the addition of Nb enhanced the milling duration for the formation of full FeSiBAlNiNb amorphous phase and also decreased the glass forming ability. Moreover, the resultant alloy has higher thermal stability and heat resistance properties. They reported that as milled FeSiBAlNiNb powders have low coercivity and, hence behave as a soft-magnetic material. As the milling duration increases Ms decreased and become lowest when the amorphous HEAs are formed. It is found that as milled products with solid solution phase having better soft magnetic characteristics as compared to fully amorphous phases. Moreover, the addition of Nb does not improve the soft-magnetic characteristics FeSiBAlNi HEA. Also both the amorphous HEAs have similar soft magnetic characteristics after prolong milling.

Xu et al. also investigate the effect of addition of C and Ce on magnetic characteristics of FeSiBAlNi (based-W5) HEAs [30]. Amorphous HEAs has been synthesized through MA. They found that addition of Ce reduced the MA time for the formation of amorphous phase and hence enhanced the glass forming ability (GFA) of the parent HEA. With the addition of C in FeSiBAlNi the formation of amorphous phase and a small amount of Si nanocrystals has been found by the authors. By the addition of both(C and Ce) the thermal stability of the alloy is enhanced. They found that with addition of C the value of Ms increased appreciably. However, with the addition of Ce the value of Ms decreased. This happened because, the addition of Ce promoted the formation of single BCC phase at initial milling time, while for the case of parent alloy BCC and FCC phases was formed. It is well known that BCC phase having lower atomic packing density as compared to FCC solid solution and hence having less ferromagnetic element and a lower total magnetic moment per unit volume as compared to FCC solid solution. Hence yield lower value of Ms with the addition of Ce in base alloy [20]. Moreover the value of Ms decreased as the milling duration increased, because remarkable change may occur in the magnetic moment through the change of the neighbor configuration of the ferromagnetic elements. The decrease in Ms may also be related to the enhanced density of grain boundaries and to the volume fraction of amorphous phase. For the synthesized samples, Hc has been found in the range of 50–378 Oe, indicating that the studied samples having semi-hard magnetic property. Different composition and microstructure together with lattice distortion due to the presence of C and Ce inevitably affect the magnetic domain wall and hence affect the Hc. Moreover, the particle size of FeSiBAlNiCe is less as compared to FeSiBAlNiC also responsible for the higher Hc of FeSiBAlNiC as compared to FeSiBAlNiCe HEA. While constant Hc has been reported after 140 h MA suggested due to the stable particle size and proportion of amorphous and solid solution phase with prolong milling duration.

Xu et al. studied the effects of addition Co, Cu and Ag on the microstructure, thermal property and magnetic properties of mechanically alloyed FeSiBAlNi HEAs [31]. They reported that as milled HEAs showed semi hard magnetic characteristics. Moreover, the value of Ms decreased with the addition of elements in base alloy. They concluded that the coexistence of FCC and BCC phases is beneficial to give rise to the semi hard magnetic characteristics than the fully amorphous or mixture of amorphous plus FCC phase. Further, Zhu et al. investigated the effect of annealing on microstructure thermal stability and magnetic characteristics of MA FeSiBAlNiM (M = Co, Cu, Ag) amorphous high entropy alloys (HEAs) [32]. High phase stability during heating process has been observed with the addition of Co and Ag in FeSiBAlNi. They found that at high annealing temperature HEAs possessed better semi-hard magnetic characteristics. They also reported that the formation of FeSi-rich and FeB-rich phases after annealing is beneficial for the enhanced magnetic characteristics of annealed HEAs. Moreover, they reported that for FeSiBAlNiCo HEAs annealing near re-crystallization temperatures was beneficiatial for the semi-hard magnetic characteristics.

Wei et al. investigated the effect of cooling rate on phase formation and magnetic characteristics of $Fe_{26.7}Co_{28.5}Ni_{28.5}Si_{4.6}B_{8.7}P_3$ HEA [33]. They found amorphous phase was formed with melt spinning at high cooling rate, whereas FCC solid solution phase has been formed at low cooling rate. The better soft magnetic characteristics have been reported for amorphous phase (saturation magnetization (Bs) of 1.07 T and coercivity (Hc) of 4 A/m) as compared to solid solution phase (Bs of 1.0 T and Hc of 168 A/m). They concluded that Bs is more sensitive towards the composition and atomic level structure and less dependent on the grain size [20]. As compared to amorphous alloy without symmetry, the low Bs has been found for the solid solution phase because high symmetry of FCC phase neutralized the atomic magnetic moment as compared to the amorphous phase. Moreover, Hc is sensitive to grain size, larger grain having high Hc [20, 34, 35].

Zuo et al. prepared equiatomic CoFeMnNi based HEA and investigated the effect of addition of non-magnetic element (Al, Cr, Ga and Sn) on crystal structure and magnetic characteristics [36]. They pointed out that for CoFeMnNi base alloy the value of Ms and Hc are found to be 18.14 emu/g and 119.9 A/m respectively. The addition of Al, Ga and Sn in CoFeMnNi separately increased the saturation magnetization from 18.14 to 147.86, 80.43 and 80.29 emu/g respectively. Furthermore, addition of Cr in base alloy dramatically reduced the value of Ms (1.39 emu/g) and enhanced the Hc up to 10,804 A/m. They reported that the apparent enhancement in Ms of FeCoMnNiX (Al, Ga and Sn) HEAs is associated to the dramatic change in crystal structure, which accordingly influenced the electronic and magnetic structures. They assumed that the antiferromagnetism of Mn atoms in CoFeMnNi base alloy is suppressed to favor the ferromagnetism due to the doping of Al, Ga and Sn. Further, Na et al. substituted the Mn element to Cr and studied the ferromagnetic transition and magnetic behaviour in equiatomic FeCoNiCrX (X = Al, Ga, Mn and Sn) HEAs [37]. They reported that FeCoNiCrMn and FeCoNiCrSn exhibited paramagnetic behaviour. While, addition of Al and Ga in FeCoNiCr base alloy changed the magnetic state from paramagnetic to ferromagnetic. Further, it has been found that with the addition of Al and Ga in base alloy, the value of Ms and Tc increased from 0.5 (Tc = 104 K) to 25 emu/g (Tc = 277 K) and 38 emu/g (Tc = 703 K) respectively. They pointed out that the enhancement in Ms is associated to the partial phase transition (formation of BCC phase).

3. Effects of compositional variation on magnetic characteristics of different HEAs

Apart from the elemental effect on magnetic characteristics of HEAs, the composition of alloying elements also play an important role on magnetic properties of the HEAs [19]. Wang et al. investigated the effect of addition of Ti on microstructure, mechanical property and magnetic characteristics of $CoCrCuFeNiTi_x$ (x = 0, 0.5, 0.8 and 1.0) HEA [38]. Both CoCrCuFeNi and $CoCrCuFeNiTi_{0.5}$ HEAs form a single FCC structure, while as the content of Ti increased in the base alloy, an intermetallic Laves phase of Fe_2Ti type has been evolved along with FCC. They pointed out that both CoCrCuFeNi and $CoCrCuFeNiTi_{0.5}$ HEAs have low value of Ms and exhibits paramagnetic nature. However, $CoCrCuFeNiTi_{0.8}$ and CoCrCuFeNiTi have superparamagnetic nature with the value of Ms of 1.36 and 1.51 emu/g respectively. The superparamagnetic nature of these HEAs is due to the nanoprecipitation.

Zhang et al. substituted Cu from Al and studied the effect of addition of different Al concentration and annealing on mechanical, electrical and magnetic properties of $CoCrFeNiTiAl_x$

(x = 0, 0.5, 1.0, 1.5 and 2.0) HEA [39]. They reported that as the content of Al increased, the value of electrical resistivity of as cast HEAs varied from 107 to 60 μΩ cm. All the as cast and as annealed HEAs exhibit high electrical resistivity as compared to the relative pure elements. It has been found that all annealed HEAs exhibit high electrical resistivity than the as cast HEAs. The enhancement in electrical resistivity after annealing has been associated to the evolution of phase composition and lattice distortion. Zhang et al. also pointed out that the saturation magnetization of the as cast $CoCrFeNiTiAl_x$ HEAs firstly increased as the content of Al increased from x = 0 to 1.0 and then decreased for x = 2.0. The as cast $CoCrFeNiTiAl_{1.0}$ HEA exhibits the best magnetic characteristics having high Ms (14.75 emu/g) with low coercivity (Hc = 22 Oe) than the other as cast HEAs. However, after vacuum annealing at 1000°C for 2 h the value of Ms decreased as compared to their as cast ones. While, the value of Ms for $CoCrFeNiTiAl_{2.0}$ HEA increased from 0.76 (as cast) to 15.74 emu/g and exhibited better magnetic characteristics as compared to the as cast and as annealed $CoCrFeNiTiAl_x$ HEAs. The value of Hc of annealed HEAs lies in the range of 20–285 Oe. They concluded that the as cast and as annealed $CoCrFeNiTiAl_x$ HEAs exhibited semi hard magnetic characteristics and these materials can be used in magnetic relayers, magnetic hysteresis motors and signal memory devices.

Kao et al. investigated the electrical and magnetic properties of Cu and Ti free $CoCrFeNiAl_x$ (x = 0–2.0) HEAs at various temperature range from 4.2 to 300 K [24]. It has been found that as the content of Al increased the structure of the alloy changed from a single FCC to a mixture of FCC and BCC and then it transformed into a single BCC phase. They studied the magnetic properties of these alloys in homogenized condition (denoted as H-x) at three different temperatures. All the homogenized alloys (H-x) exhibit ferromagnetic nature at low temperature (5 and 50 K). In addition to this the value of Ms for H-1.25 and H-2.0 HEAs exceeds at low temperature than that of the H-0 and H-0.25 alloys. This indicated that at low temperature, the BCC phase has a higher value of Ms than the FCC phase. At room temperature (300 K), the homogenized HEAs H-0, H-0.25 and H-0.75 exhibit paramagnetic behaviour, while H-0.5, H-1.25 and H-2.0 remain ferromagnetic. Kao et al. calculated the phase contribution of both FCC and BCC phases for H-0.50 and H-0.75 HEAs at 5 K. It has been found that the value of Ms for FCC phase is higher than the BCC phase. The reason for $Ms_{BCC} < Ms_{FCC}$ is associated to the existence of Al or AlNi-rich phase. The value of Ms decreased in the range of $0 \leq x \leq 0.25$ and $1.25 \leq x \leq 2.00$. The value of Ms for H-0.25 alloy is smaller than that of the H-0 HEA, indicating that the addition of Al reduces the magnetic moment. Besides this, H-2.0 alloy has smaller value of Ms than the H-1.25 HEA. Kao et el. pointed out that this is due to the high content of ordered BCC phase. It has been observed that H-2.00 alloy is mainly composed of AlNi-rich ordered BCC phase, whereas the H-1.25 alloy has less content of AlNi-rich ordered BCC phase. They concluded that the addition of Al reduced the ferromagnetic nature of single FCC and single BCC H-x alloys. In single BCC phase the reduction in the value of Ms is associated to the high content of AlNi-rich ordered BCC phase.

In a similar manner, Vrtink et al. also investigated the magnetic behaviour of $CoCrFeNiZr_x$ (x = 0.40, 0.45 and 0.50) eutectic HEAs at different temperatures (5, 100 and 300 K) [40]. They reported that two magnetic structures, a disordered ferromagnetic (F) and a superparamagnetic-like (S) coexist in the $CoCrFeNiZr_x$ HEAs. The $CoCrFeNiZr_{x\ =\ 0.40,0.45,0.50}$ HEAs exhibit the paramagnetism at 300 K, whereas at low temperature (5 and 100 K) it shows ferromagnetism. In addition to this, they also reported that at low temperature $CoCrFeNiZr_{0.45}$ and $CoCrFeNiZr_{0.50}$ HEA exhibit negligible coercive field as compared to the $CoCrFeNiZr_{0.40}$. Vrtinke et al. explained that this difference in the coercive field is associated to the vanishing of disordered F phase.

Ma et al. investigated the effect of Nb addition on the microstructure, mechanical property and magnetic behaviour of $AlCoCrFeNiNb_x$ (x = 0, 0.1, 0.25, 0.5 and 0.75) HEAs [41]. They reported that as the content of Nb increased a single BCC phase has been transformed into a mixture of (CoCr)Nb type Laves phase and BCC phase. Additionally, the value of Ms decreased and Hc increased from 64 to 10.31 emu/g and 52 to 94 Oe respectively as content of Nb increased. They reported that the alternation in the value of Ms and Hc may be associated to the magnetic hardening of the CoCrNb alloy and the Laves phase may pin the magnetic domain of BCC phase.

Recently, Zhang et al. reported a new class of CoFeNi based $CoFeNi(AlSi)_x$ ($0 \leq x \leq 0.8$) HEAs and investigated their structural, mechanical, magnetic and electrical behaviour [20]. They reported that the value of saturation magnetization is primarily determined by the composition and atomic level structures. With increasing the content of Al and Si, the value of Ms monotonically decreased from 1.315 to 0.46 T. Different from saturation magnetization it has been found that the coercivity is highly dependent on the microstructure, grain size and lattice distortion. They reported that addition of Al and Si induced the lattice distortion and changes the microstructure of the HEAs that inevitably affects the magnetic domain wall movement and, thus, the coercivity. It has been found that the optimal balance of magnetic, electrical, and mechanical properties is achieved at x = 0.20, which has the best combination of Ms (1.15 T), Hc (1400 A/m), resistivity ($\rho = 69.5\ \mu\Omega$ cm), yield strength (342 MPA) and strain without fracture ($\varepsilon_p > 50\%$) for application as soft magnetic materials.

In a similar manner to Zhang et al. [20], the effects of phase constitution on magnetic and mechanical properties of $FeCoNi(CuAl)_x$ (x = 0–1.2, in molar ratios) HEAs has been investigated by the Zhang et al. [25]. They reported that depending on the content of alloying elements (Cu and Al), the $CoFeNi(CuAl)_x$ HEAs exhibit single FCC for $0 \leq x \leq 0.6$, BCC combined with minor FCC for $0.9 \leq x \leq 1.2$ and duplex FCC + BCC phase in the range of $0.7 \leq x < 0.9$. The value of Ms decreased as the content of Cu and Al increased from x = 0 to x = 1.2, whereas some deviation in the range of $0.8 \leq x \leq 0.9$ has been observed. They reposted that the value of Ms is highly sensitive to the composition and phase constitution than their microstructure. Additionally, they annealed these as cast HEAs in three different temperatures (573–673 K) and they found that heat treatment did not impose any impacts on the magnetic characteristics for single FCC phase ($0 \leq x \leq 0.6$) and nearly single BCC phase ($0.9 \leq x \leq 1.2$) HEAs. However, the value of Ms increased for FCC + BCC duplex phase ($0.7 \leq x < 0.9$ HEAs. For $CoFeNi(CuAl)_{0.8}$ HEA, the value of Ms increased from 78.9 (as cast) to 93.1 emu/g (annealed at 673 K). They pointed that the enhancement in the value of Ms after annealing is associated to the phase transition from FCC to BCC for $CoFeNi(CuAl)_{0.8}$ HEA. In summery they concluded that through the proper alloy composition and annealing one can achieve a superior soft magnetic metallic alloys with high saturation magnetization and high ductility.

Lie et al. also studied the composition dependence of crystal structure, physical and mechanical properties of $FeCoNi(MnAl)_x$ (x = 0, 0.25, 0.5, 0.75, 1) high entropy alloys [42]. They reported that as the value of x increased from x = 0 to x = 1, the crystal structure changed from single FCC ($0 \le x \le 0.25$) to FCC + BCC ($0.5 \le x \le 0.75$) and then to a single BCC phase (x = 1). They also reported that with the variation of x values, the value of Ms firstly decreased from FeCoNi (155.7 emu/g) to $FeCoNiMn_{0.5}Al_{0.5}$ (51.9 emu/g) and increased from $FeCoNiMn_{0.75}Al_{0.75}$ 129.6 emu/g) to FeCoNiMnAl (132.2 emu/g). In addition to the saturation magnetization the value of Hc firstly increased from FeCoNi (189 A/m) to $FeCoNiMn_{0.5}Al_{0.5}$ (730 A/m) and decreased from $FeCoNiMn_{0.75}Al0_{0.75}$ (445 A/m) to FeCoNiMnAl (266 A/m) with the variation of x. They reported that the alternation in the value of Ms is associated to the change in lattice parameter and crystal structure. While the alternation in coercivity is associated to the change in microstructure, together with the lattice distortion induced by Al atoms. Moreover, Lie et al. also studied the magnetic characteristics of cold rolled and annealed FeCoNi, $FeCoNiMn_{0.25}Al_{0.25}$ and $FeCoNiMn_{0.5}Al_{0.5}$ alloys. They concluded that annealed FeCoNi and $FeCoNiMn_{0.25}Al_{0.25}$ alloys exhibit the optimal balance of magnetic and mechanical behaviour.

We have also investigated the effect of elemental addition (Mn and Co) on the phase formation and magnetic characteristics of TiFeNiCr base HEAs [15]. HEAs with different elements have been synthesized through the mechanical alloying. XRD analysis of synthesized HEAs confirmed that double FCC and a minor sigma phase have been evolved for TiFeNiCr and TiFeNiCrMn HEAs. However, single FCC phase has been appeared for TiFeNiCrCo HEA. The magnetic characteristic of synthesized and annealed HEA investigated through the VSM at room temperature. The value of Ms for TiFeNiCr HEA was found to be 13.82 emu/g (as shown in **Figure 1(a)**). However, the value of Ms for TiFeNiCrMn and TiFeNiCrCo HEAs was found to be 2.28 and 24.44 emu/g respectively (as shown in **Figure 1(b)** and **(c)**). It is concluded

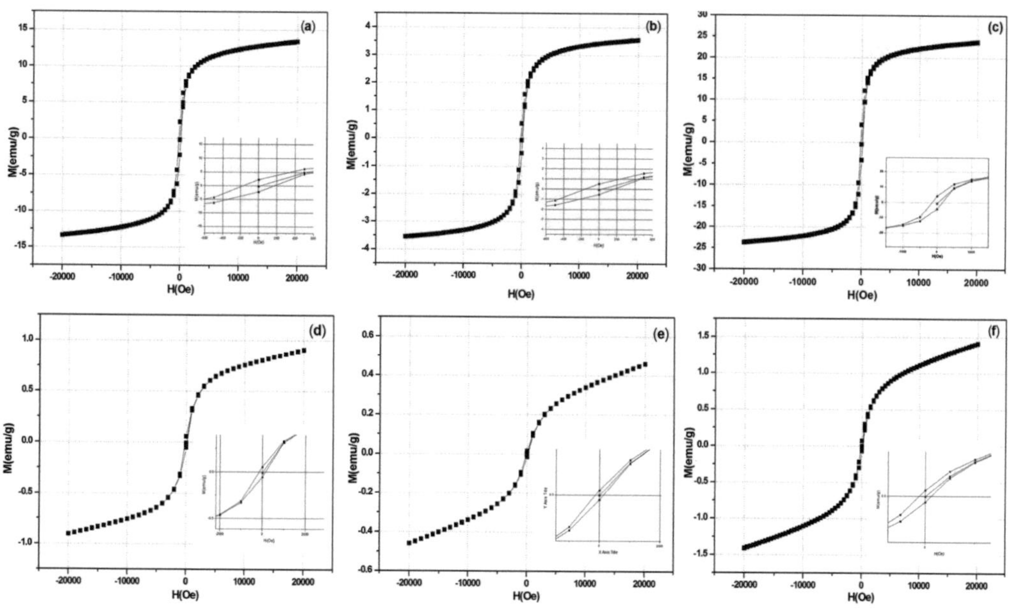

Figure 1. Magnetic hysteresis curve of (a) TiFeNiCr, (b) TiFeNiCrMn, (c) TiFeNiCrCo as synthesized and (d) TiFeNiCr, (e) TiFeNiCrMn, (f) TiFeNiCrCo annealed at 700°C HEAs. The inset represents the magnified view of selected region.

from the M-H curve, with the addition of ferromagnetic element (like Co) the value of Ms increased, while with addition of paramagnetic element (like Mn) the value of Ms decreased appreciably. Thus among TiFeNiCr, TiFeNiCrCo and TiFeNiCrMn HEAs the value of Ms was highest for the case of TiFeNiCrCo HEA because this alloy contained three ferromagnetic elements, i.e., 60 at.% (Fe, Ni and Co) of the alloy is ferromagnetic in nature. Thus, the value of Ms depends on the presence of magnetic elements and as the content of ferromagnetic elements increased the value of Ms increased appreciably. The value of Hc for TiFeNiCr, TiFeNiCrMn and TiFeNiCrCo HEAs was found to be 166.93, 225.83 and 149.54 Oe. We have also been investigated the effect of annealing on magnetic characteristics of HEAs. From figure it can say that the value of Ms, M_R/M_s and Hc decreased dramatically after annealing of the synthesized HEAs at 700°C (shown in **Figure 1**). The reduction in Ms after annealing may be attributed to the increase in lattice parameter after annealing. It has been found that the lattice parameter after annealing increased from 3.592 to 3.595 Å, 3.575 to 3.583 Å and 3.561 to 3.582 Å for TiFeNiCr, TiFeNiCrMn and TiFeNiCrCo, respectively. Hence the separation between the ferromagnetic elements of HEAs increased after annealing due to which the magnetic exchange coupling is altered and the value of Ms decreased for annealed TiFeNiCr, TiFeNiCrMn and TiFeNiCrCo HEAs.

4. Conclusions

High entropy alloys having multi-principal elements and has growing interest of Scientist Engineers and Metallurgists. The applications of HEAs are based on their four core effects. Among these effects the cocktail effect is the most effective for the functional properties of these alloys. HEAs may have wide range of applications along with better functional and structural properties. They have excellent mechanical properties. A soft magnetic material should have good mechanical property, structural stability at high temperature and low coercivity with high magnetization. Recently, reported $FeCoNiMn_{0.25}Al_{0.25}$ and CoCrFeNiM (M = Cu, Mn) HEAs got attention as a better soft magnetic material because these HEAs having good soft magnetic characteristics along with good mechanical and excellent structural stability at high-temperature. We have summarized the key results of magnetic characteristics of some recently investigated promising high entropy alloys. Further the mechanical as well as the magnetic characteristics of HEAs can be tuned through the selection of proper alloying elements and synthesis route. The magnetic characteristics of these alloys basically depend on the alloying element and compositional variation of the magnetic element present in particular HEAs. Thus the magnetic as well as mechanical properties of these alloys can be tuned based on our requirement/applications. As can be seen from the summarized results by the variation of single element the magnetic nature of the material can be changed to ferromagnetic to paramagnetic or super paramagnetic. Similarly, hard, semi-hard and soft magnetic characteristics of the materials have been changed by the change in the processing route of the HEAs. Thus we can conclude that alloying elements, composition and processing route has significant effect on phase evolution and microstructure, which inevitably affects the magnetic characteristics of HEAs. Among them, some HEAs (e.g $FeCoNiAl_{0.2}Si_{0.2}$, $FeCoNiMn_{0.25}Al_{0.25}$) have been proven to have better soft magnetic characteristics with high Curie temperature (Tc) and excellent mechanical behavior as compared to conventional alloys. Therefore, more researches are needed in future for design and development of new HEAs and to explore their properties for application-oriented viewpoints.

Acknowledgements

The author would like to acknowledge Prof. O.N. Srivastava (BHU) for the scientific discussions and encouragement. Financial assistance received from DST INSPIRE Faculty Research Grant (IFA-12-PH43), New Delhi (India) is gratefully acknowledged.

Conflict of interest

Authors declare that this chapter has no conflict of interest.

Author details

Rajesh K. Mishra and Rohit R. Shahi*

*Address all correspondence to: rohitrshahi@gmail.com

Department of Physics, Motilal Nehru National Institute of Technology Allahabad, Allahabad, India

References

[1] Yeh JW, Chen SK, Lin SJ, Gan JY, Chin TS, Shun TT, Tsau CH, Chang SY. Nanostructured high entropy alloys with multiple elements: Novel alloy design concept and outcomes. Advanced Engineering Materials. 2004;**6**:299-303

[2] Tsai MH, Yeh JW. High-entropy alloy: A critical review. Materials Research Letters. 2014; **2**(3):107-123

[3] Zhang Y, Zuo TT, Tang Z, Gao MC, Dahmen KA, Liaw PK, Lu ZP. Microstructures and properties of high entropy alloys. Progress in Materials Science. 2014;**61**:1-93

[4] Miracle DB, Senkov ON. A critical review of high entropy alloys and related concepts. Acta Materialia. 2017;**122**:448-511

[5] Liu ZY, Guo S, Liu XJ, Ye JC, Yang Y, Wang XL, Yang L, An K, Liu CT. Micromechanical characterization of casting-induced inhomogeneity in an $Al_{0.8}CoCrCuFeNi$ high-entropy alloy. Scripta Materialia. 2011;**64**:868-871

[6] Fu ZQ, Chen WP, Wen HM, Zhang DL, Chen Z, Zheng BL, Zhou YZ, Lavernia EJ. Microstructure and strengthening mechanisms in an FCC structured single-phase nanocrystalline $Co_{25}Ni_{25}Fe_{25}Al_{7.5}Cu_{17.5}$ high entropy alloy. Acta Materialia. 2016;**107**:59-71

[7] Otto F, Yang Y, Bei H, George EP. Relative effects of enthalpy and entropy on the phase stability of equiatomic high-entropy alloys. Acta Materialia. 2013;**106**:2628-2638

[8] Guo S, Liu CT. Phase stability in high entropy alloys: Formation of solid solution or amorphous phase. Progress in Natural Science: Materials International. 2011;**21**:433-446

[9] Guo S, Ng C, Lu J, Li CT. Effect of valence electron concentration on stability of FCC or BCC phase in high entropy alloys. Journal of Applied Physics. 2011;**109**:103505

[10] Alaneme KK, Bodunrin MO, Oke SR. Processing, alloy composition and phase transition effect on the mechanical and corrosion properties of high entropy alloys: A review. Journal of Materials Research and Technology. 2016;**5**:384-393

[11] Murti BS, Yeh JW, Ranganathan S. High Entropy Alloys. 3rd ed. London: Elsevier; 2014

[12] Gao MC, Yeh JW, Liaw PK, Zhang Y. High Entropy Alloys: Fundamental and Application. 1st ed. Switzerland: Springer; 2017

[13] Ghazi SS, Ravi KR. Phase-evolution in high entropy alloys: Role of synthesis route. Intermetallics. 2016;**73**:40-42

[14] Zhang K, Fu Z. Effects of annealing treatment on phase composition and microstructure of CoCrFeNiTiAlx high entropy alloys. Intermetallics. 2012;**22**:24-32

[15] Mishra RK, Shahi RR. Phase evolution and magnetic characteristics of TiFeNiCr and TiFeNiCrM(M=Mn, Co) high entropy alloys. Journal of Magnetism and Magnetic Materials. 2017;**442**:218-223

[16] Shun TT, Chang LY, Shiu MH. Microstructures and mechanical properties of multiprincipal component CoCrFeNiTix alloys. Materials Science and Engineering A. 2012;**556**:170-174

[17] Shun TT, Chang LY, Shiu MH. Microstructure and mechanical properties of multiprincipal component CoCrFeNiMox alloys. Materials Characterization. 2012;**70**:63-67

[18] Liu WH, He JY, Huang HL, Wang H, Lu ZP, Liu CT. Effects of Nb additions on the microstructure and mechanical property of CoCrFeNi high-entropy alloys. Intermetallics. 2015;**60**:1-8

[19] Tsai MH. Physical properties of high entropy alloys. Entropy. 2013;**15**:5338-5345

[20] Zhang Y, Zuo TT, Cheng YQ, Liaw PK. High entropy alloys with high saturation magnetization, electrical resistivity and malleability. Scientific Reports. 2013;**3**:1455

[21] Yu PF, Zhang LJ, Chen H, Zhang H, Ma MZ, Li YC, Li G, Liaw PK. The high entropy alloys with high hardness and soft magnetic property prepared by mechanical alloying and high pressure sintering. Intermetallics. 2016;**17**:82-87

[22] Li P, Wang A, Liu CT. A ductile high entropy alloy with attractive magnetic properties. Journal of Alloys and Compounds. 2017;**694**:55-60

[23] Zuo TT, Li RB, Ren XJ, Zhang Y. Effects of Al and Si addition on the structure and properties of CoFeNi equal atomic ratio alloy. Journal of Magnetism and Magnetic Materials. 2014;**371**:60-68

[24] Kao YF, Chen SK, Chen TJ, Chu PC, Yeh JW, Lin SJ. Electrical, magnetic, and Hall properties of AlxCoCrFeNi high-entropy alloys. Journal of Alloys and Compounds. 2011;**509**:1607-1614

[25] Zhang Q, Xu H, Tan XH, Hou XL, Wu SW, Tan GS, Yu LY. The effects of phase constitution on magnetic and mechanical properties of FeCoNi(CuAl)x (x = 0–1.2) high-entropy alloys. Journal of Alloys and Compounds. 2017;**693**:1061-1067

[26] Lucas MS, Mauger L, Munoz JA, Xiao Y, Sheets AO, Semiatin SL, Howath J, Turgut Z. Magnetic and vibrational properties of high entropy alloys. Journal of Applied Physics. 2011;**307**:07E109

[27] Zhang KB, Fu ZY, Zhang JY, Shi J, Wang WM, Wang H, Wang YC, Zhang QJ. Annealing on the structure and properties evolution of the CoCrFeNiCuAl high-entropy alloy. Journal of Alloys and Compounds. 2010;**502**:295-299

[28] Singh S, Wanderka N, Kiefer K, Siemensmeyer K, Banhart J. Effect of decomposition of the Cr–Fe–Co rich phase of AlCoCrCuFeNi high entropy alloy on magnetic properties. Ultramicroscopy. 2011;**111**:619-622

[29] Wang J, Zheng Z, Xu J, Wang Y. Microstructure and magnetic properties of mechanically alloyed FeSiBAlNi (Nb) high entropy alloys. Journal of Magnetism and Magnetic Materials. 2014;**355**:58-64

[30] Xu J, Axinte E, Zhao Z, Wang Y. Effect of C and Ce addition on the microstructure and magnetic property of the mechanically alloyed FeSiBAlNi high entropy alloys. Journal of Magnetism and Magnetic Materials. 2016;**414**:59-68

[31] Xu J, Shang C, Ge W, Jia H, Liaw PK, Wang Y. Effects of elemental addition on the microstructure, thermal stability and magnetic properties of the mechanically alloyed FeSiBAlNi high entropy alloys. Advanced Powder Technology. 2016;**27**:1418-1426

[32] Zhu X, Zhou X, Yu S, Wei C, Xu J, Wang Y. Effects of annealing on the microstructure and magnetic property of the mechanically alloyed FeSiBAlNiM (M = Co, Cu, Ag) amorphous high entropy alloys. Journal of Magnetism and Magnetic Materials. 2017;**430**:59-64

[33] Wei R, Sun H, Chen C, Han Z, Li F. Effect of cooling rate on the phase structure and magnetic properties of $Fe_{26.7}Co_{28.5}Ni_{28.5}Si_{4.6}B_{8.7}P_3$ high entropy alloy. Journal of Magnetism and Magnetic Materials. 2017;**435**:184-186

[34] Anand G, Goodall R, Freeman CL. Role of configurational entropy in body-centred cubic or face-centred cubic phase formation in high entropy alloys. Scripta Materialia. 2016;**124**: 90-94

[35] Herzer G. Modern soft magnets: Amorphous and nanocrystalline materials. Acta Materialia. 2013;**61**:718-734

[36] Zuo TT, Gao MC, Ouyang L, Yang X, Cheng YQ, Feng R, Chen S, Liaw PK, Hawk JA, Zhang Y. Tailoring magnetic behavior of CoFeMnNiX (X = Al, Cr, Ga, and Sn) high entropy alloys by metal doping. Acta Materialia. 2017;**130**:10-18

[37] Na SM, Yoo JH, Lambert PK, Jones NJ. Room-temperature ferromagnetic transitions and the temperature dependence of magnetic behaviors in FeCoNiCr-based high-entropy alloys. AIP Advances. 2018;**8**:056412

[38] Wang XF, Zhang Y, Qiao Y, Chen GL. Novel microstructure and properties of multicomponent CoCrCuFeNiTix alloys. Intermetallics. 2007;**15**:357-362

[39] Zhang K, Fu Z. Effects of annealing treatment on properties of CoCrFeNiTiAlx multi-component alloys. Intermetallics. 2012;**28**:34-39

[40] Vrtnik S, Guo S, Sheikh S, Jelen A, Koželj P, Luzar J, Kocjan A, Jagličić Z, Meden A, Guime H, Kime HJ, Dolinšek J. Magnetism of CoCrFeNiZrx eutectic high entropy alloys. Intermetallics. 2018;**93**:122-133

[41] Ma SG, Zhang Y. Effect of Nb addition on the microstructure and properties of AlCoCrFeNi high-entropy alloy. Materials Science and Engineering A. 2012;**532**:480-486

[42] Li P, Wang A, Liu CT. Composition dependence of structure, physical and mechanical properties of FeCoNi(MnAl)x high entropy alloys. Intermetallics. 2017;**87**:21-26

Section 4

Magnetic Modelling

Chapter 5

Mathematical Model on Magnetic Drug Targeting in Microvessel

Sachin Shaw

Additional information is available at the end of the chapter

http://dx.doi.org/10.5772/intechopen.73678

Abstract

Drug targeting is a process by which the distribution of drug in an organism is deployed in such a manner that its major fraction interacts exclusively with the target tissue at the cellular or subcellular level. Magnetic drug targeting is one of the major drug delivery methods due to its noninvasiveness, high targeting efficiency, and minimized toxic side effects on healthy cells and tissues. There are several experimental works on the magnetic drug targeting through microvessel, but very few works are carried out on the mathematical models on magnetic drug delivery. The aim of the present chapter is to discuss all major and minor factors, such as fluidic force, magnetic force, particle-particle interaction, inertia force, Saffman lift force, permeability of the microvessel and carrier particle, and so on, which influenced the drug targeting through microvessel by considering the nature of blood flow as Newtonian, non-Newtonian, single phase, and two phase model. A brief details of fluidic force, magnetic force, particle-particle interaction, Saffman force, buoyancy force, etc. Mathematical models on the fluidic force are discussed for Newtonian, non-Newtonian fluid, single phase, and two-phase fluid model including other forces that influence the magnetic drug targeting in microvessel.

Keywords: magnetic drug targeting, mathematical model, non-Newtonian fluid, single phase, two-phase flow

1. Introduction

Magnetic micro- and nanoparticles are finding increasing use in different fields of microbiology, biomedicine, and biotechnology where they are used to transport and separate materials, label, and to deliver therapeutic drugs to a target tissue. The use of the magnetic particles as transport agents in different bio-applications has been discussed by Furlani [1]. They have discussed the advantages of using magnetic nanoparticles which are well suited for the different bio-applications due to the following reasons: (1) they are nontoxic and well tolerated by living organisms after being well synthesized and functionalized; (2) they can be synthesized in sizes that range from a few nanometers to higher ranges with a very narrow interval. This nature makes them ideal for probing and manipulating bioparticles and biosystems of different ranges such as protein (5–50 nm), viruses (20–50 nm), genes (2 nm wide and 10–100 nm long), or whole cells (10–100 μm). It is also noted that sub-micron and micron-sized magnetic nanoparticles are used in different bio-applications; (3) magnetic nanoparticles are very familiar with custom-tailored surface treatment to enhance biocompatibility and enable coating with affinity biomolecules for highly specific binding with a target biomaterial; (4) magnetic nanoparticles can easily be magnetized by an applied magnetic field, but once field is removed, it reverts back to an unmagnetized state. This behavior can easily be used to separate or immobilize magnetically labeled biomaterials from a carrier fluid using an external magnetic field. Significantly, the relatively low permeability of an aqueous carrier fluid enables efficient magnet coupling to an immersed magnetically labeled biomaterial. Moreover, the low intrinsic magnetic susceptibility of most biomaterials provides substantial contrast between labeled and unlabeled material, which enables a high degree of selectivity and detection. Magnetic labeling has advantages over conventional fluorescence and chemiluminescence-based biolabels. Notably, small samples of magnetically labeled material can be detected using ultra-sensitive ferromagnetic "spin valve" sensors, which can be integrated into microfluidic-based diagnostic systems.

Magnetic particles have an additional advantage, mainly designed to absorb energy at a resonant frequency from a time-varying magnetic field, which enables their use for therapeutic hyperthermia of tumors. During radio frequency (RF) hyperthermia, magnetic nanoparticles are directed to malignant tissue and then irradiated with an AC magnetic field of sufficient magnitude and duration to heat the tissue to 42°C for 30 min or more, which is sufficient to destroy the tissue (Moroz et al., [2]). In the same manner, it will be used for other cancer therapies (Hergt and Duzt [3]; Gupta and Gupta [4]; Shaw and Murthy [5–7]). Magnetic nanoparticles are also used for bioimaging, both optically, using surface-bound fluorophores for biophotonic applications (Kircher et al. [8]; Sahoo et al. [9]; Sosnovik et al. [10]; Prasad [11]; Levy et al. [12]; Medarova et al. [13]) and magnetically where they serve as contrast agents for enhanced MRI.

Drug targeting is a process by which the distribution of drug in an organism is deployed in such a manner that its major fraction interacts exclusively with the target tissue at the cellular or subcellular level. Magnetic drug targeting is one of the major drug delivery methods due to its noninvasiveness, high targeting efficiency, and minimized toxic side effects on healthy cells and tissues (Lübbe et al. [14]; Alexiou et al. [15]). This drug targeting is mainly used for the medical treatment of various diseases, especially cancer, and cardiovascular and endovascular diseases, such as stenosis, thrombosis, aneurysm, atherosclerosis (Jurgons et al. [16]), trancheobronchial airways (Pourmehran et al. [17]), and so on. Magnetic drug targeting is a growing interest with recent progress in the development of carrier particles that are designed to target a specific tissue and effect local chemo-, radio- and gene therapy at the tumor site (**Figure 1**) (Fabrizio and Francois [18]; Berryl and Curtis [19]).

http://dx.doi.org/10.5772/intechopen.73678

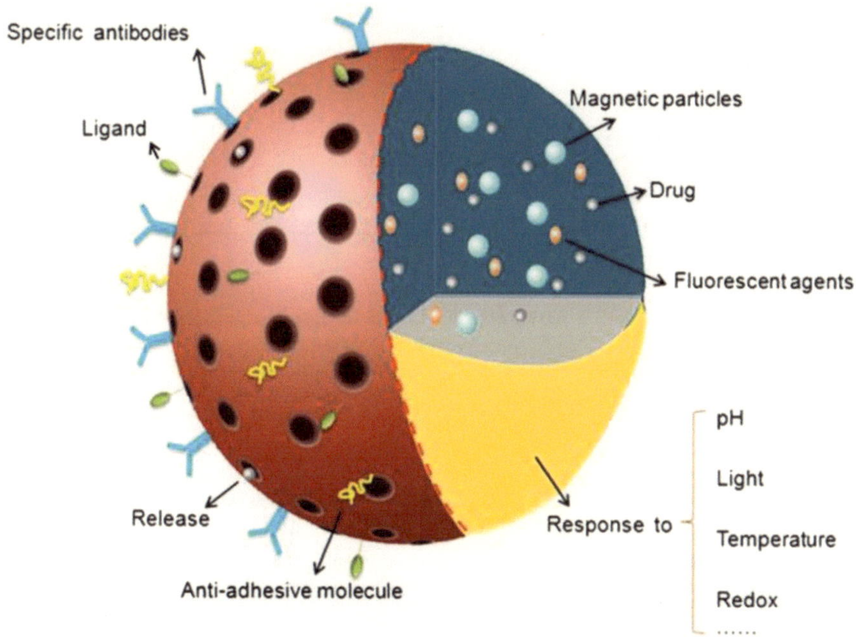

Figure 1. Schematic illustration of multifunctional and stimuli-responsive porous carrier particle, which is modified with specific antibodies, ligands targeting the cell surface, and anti-adhesive molecules (shown in the red area), and loaded with magnetic particles, fluorescent agents, and drugs for imaging, detection, and therapy (shown in the blue area). It is synthesized from stimuli-responsive polymers or modified with responsive molecules, which can respond to the change of environmental stimuli like pH, light, heat, oxides, and so on. (shown in the yellow area) (Fan et al. [20]).

Magnetic drug targeting is one of the growing interests with recent progress in the development of carrier particles that are designed to target a specific tissue to cure the tumor. In magnetic drug targeting, the therapeutic agent can be either encapsulated into a magnetic micro- or nanosphere or conjugated on its surface. Magnetic particles with bound drug molecules are injected into the vascular system upstream from the malignant tissue (**Figure 2**). They can be immobilized at the tumor site using a local magnetic field gradient produced by an external field source. Particle accumulation at the tumor is often augmented by magnetic agglomeration, and the efficiency of the accumulation depends on various physiological parameters including particle size, surface characteristics, field strength, blood flow rate, and so on (**Figure 3**).

Upon achieving a sufficient particle concentration at a tumor, drug molecules can be released from their carrier particles by changing physiological conditions such as pH, osmolality, or temperature, or by enzymatic activity (Berry and Curtis [19]; Arrueboa et al. [21]). Since the therapeutic agents are localized to regions of diseased tissue, higher dosages can be applied, which enables more effective treatment. This is in contrast to less selective conventional chemotherapy wherein a toxic drug is distributed systemically throughout the body, potentially harming healthy tissue.

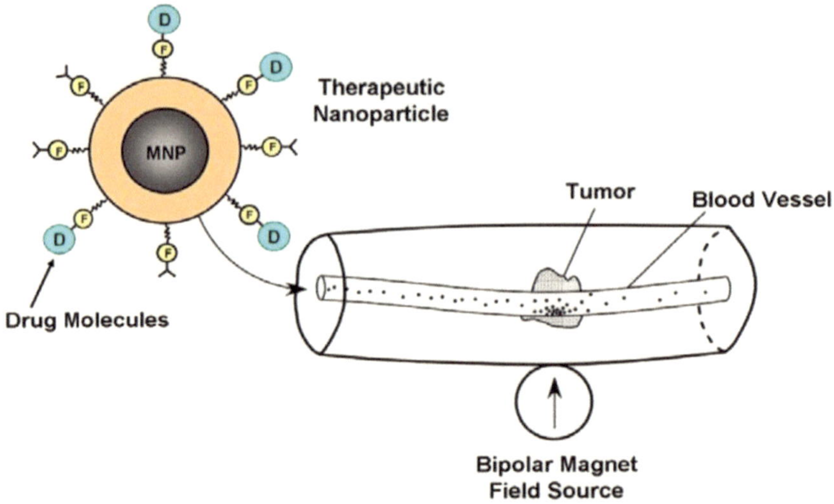

Figure 2. Noninvasive magnetic drug targeting in a microvessel (Furlani [1]).

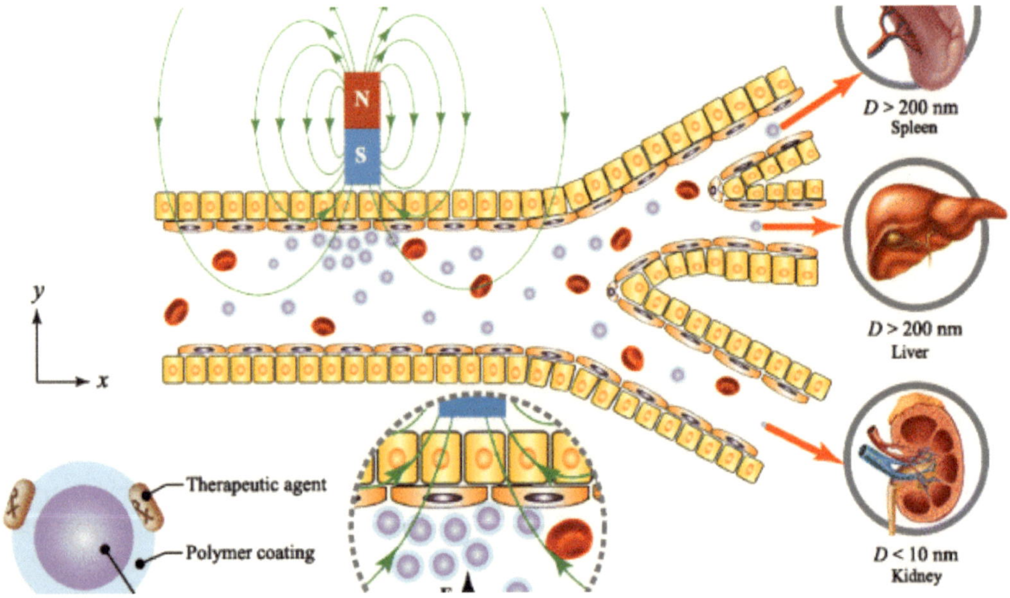

Figure 3. Drug-loaded carrier is typically composed of a magnetic core and a biocompatible coating material. The magnetic core was made from different materials such as Fe_3O_4, Fe_2O_3, or Fe. The coating materials are Au, PEG, or SiO_2. Based on the biokinetics of particles, a drug carrier ranging from 10 to 200 nm in diameter is optimal for in vivo delivery, as the small particles ($D < 10$ nm) escape by renal clearance and the large ones ($D > 200$ nm) are sequestered by the reticuloendothelial system of the spleen and liver (Lunnoo and Puangmali [22]).

During magnetic drug targeting, beyond fluidic force and magnetic force, there are several factors that play a different role in the phenomena. They are related to blood flow such as particle-particle interaction (mainly for the vessel with small radius), Saffman lift force, and so on. Some of them related to the drug particles e.g., drag force, interparticle effects such as magnetic dipole interactions, particle-blood cell interaction, etc. Here, one by one, we discuss different forces that directly influence the trajectories of the carrier particle and on magnetic drug targeting.

2. Magnetic force on the particle

Magnetic force on the particle that flows with the fluid is well described by Furlani and Furlani [23]. By linear magnetization model, relation of magnetization and intensity of the magnetic field in the saturation can be written as

$$M_{sp} = \chi_p H_{in,sat}, \tag{1}$$

where $\chi_p = \mu_p/\mu_0 - 1$ is the susceptibility of the particle, μ_p is the permeability of the particle, and μ_0 is the permeability of air. The magnetic force is determined by "effective" dipole moment method as described by Jones [24] and Furlani [25]. In this method, the magnetized particle will be replaced by an "equivalent" point dipole with a moment $m_{p,\ eff}$, and the force on the dipole (and hence on the particle) is given by

$$F_m = \mu_f \left(m_{eff}.\nabla\right)\mathbf{H}_a, \tag{2}$$

where μ_f is the permeability of the transport fluid, m_{eff} is the effective dipole moment of the particle, and H_a is the externally applied magnetic field intensity at the center of the particle, where the equivalent dipole is located. Firstly, we solve the magneto-static boundary value problem for the particle (here the considered shape of the particle is spherical, and it may change according to the shape) immersed in a fluid where the magnetization M_p is parallel to the applied field and then determine the equivalent point dipole moment $m_{p,eff}$. Spherical coordinate system $(r.\theta,\ \phi)$ is considered for the spherical particle with origin as the center of the spherical particle. Cartesian coordinates (x, y, z) are considered for the microvessel and magnet (see **Figure 4**).

The potential field inside and outside of the particle can be written as

$$\Phi_{in}(r,\theta) = \nabla \mathrm{H}_{in} = -C_{in} r\cos(\theta)\left(r < R_p\right). \tag{3}$$

$$\Phi_{out}(r,\theta) = -\nabla \mathrm{H}_{out} = -H_a r\cos(\theta) + C_{out}\,\frac{\cos(\theta)}{r^2}\left(r \geq R_p\right). \tag{4}$$

where R_p is the radius of the spherical particle. Moreover, we considered $z-$ axis in the direction of applied magnetic field. The magnitude of the field intensity inside the particle in the $z-$ axis is

$$H_{in,z} = -\left[\frac{\partial}{\partial r}\Phi_{in}(r,\theta)\right]\cos(\theta) + \frac{1}{r}\left[\frac{\partial}{\partial r}\Phi_{in}(r,\theta)\right]\sin(\theta). \tag{5}$$

The corresponding boundary condition with respect to the potential inside and outside the particle surface and normal component of resultant of the magnetic force are written as:

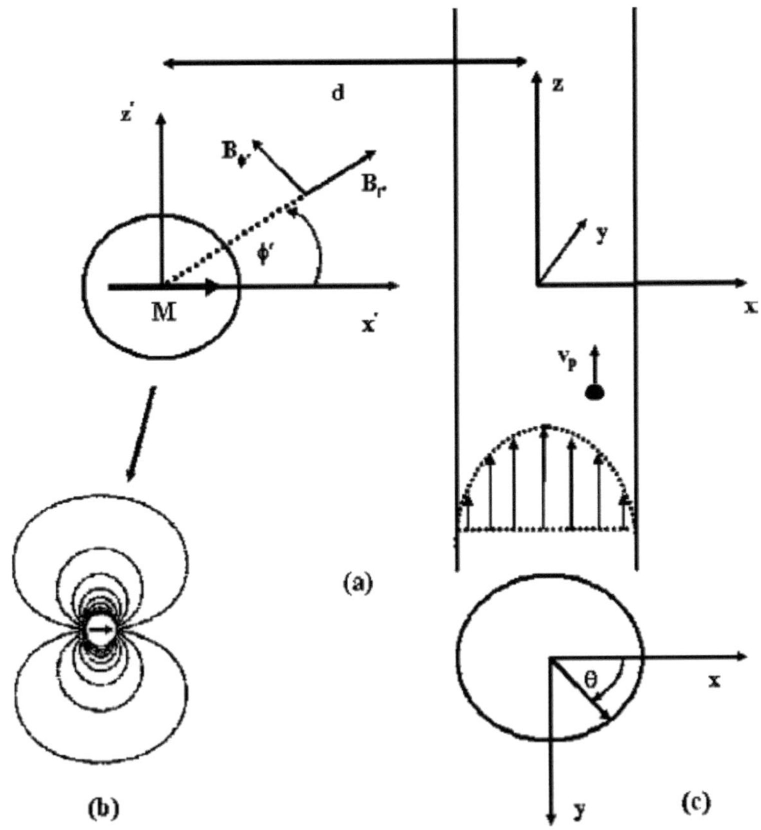

Figure 4. Geometry and reference frame for analysis: (a) coordinate systems and reference frames, (b) magnetic flux lines for a cylinder magnet, cross-section of microvessel with reference frame (Furlani [23]).

$$\Phi_{in}(r,\theta) = \Phi_{out}(r,\theta),$$
$$\mu_0\left(\frac{\partial}{\partial r}\Phi_{in} + M_p\right) = -\mu_f\frac{\partial}{\partial r}\Phi_{out}. \quad (6)$$

Solving Eqs. (3) and (4) with boundary condition (6), we get

$$C_{in} = \frac{3\mu_f}{\mu_0 + 2\mu_f}H_a - \frac{\mu_0}{\mu_0 + 2\mu_f}M_a, \quad (7)$$

and

$$C_{out} = R_p^3\left(\frac{\mu_0 - \mu_f}{\mu_0 + 2\mu_f}H_a + \frac{\mu_0}{\mu_0 + 2\mu_f}M_a\right). \quad (8)$$

From the definition (see Eq. (3)), it is clear that $H_{in} = C_{in}$ and if the particle is below the saturation, then $M_p = \chi_p H_{in}$. By substituting these two relations in Eq. (7), we get:

$$H_{in} = \frac{3\left(\chi_f + 1\right)}{\left(\chi_p - \chi_f\right) + 3\left(\chi_f + 1\right)} H_a, \tag{9}$$

and

$$M_p = \chi_p H_{in} = \frac{3\chi_p\left(\chi_f + 1\right)}{\left(\chi_p - \chi_f\right) + 3\left(\chi_f + 1\right)} H_a. \tag{10}$$

The magnitude of the "equivalent" dipole for the particle is related to C_{out}, and it can be written as

$$m_{p,eff} = 4\pi C_{out} = V_p \frac{3\left(\chi_p - \chi_f\right)}{\left(\chi_p - \chi_f\right) + 3\left(\chi_f + 1\right)} H_a, \tag{11}$$

where $V_p = (4/3)\pi R_p^3$ which is the volume of the spherical particle.

Substituting Eq. (11) into Eq. (2), we can define the magnetic force in the form of

$$F_m = \mu_f V_p \frac{3\left(\chi_p - \chi_f\right)}{\left(\chi_p - \chi_f\right) + 3\left(\chi_f + 1\right)} (\mathbf{H_a}.\nabla)\mathbf{H_a}. \tag{12}$$

Now when the susceptibility of the fluid is very small (i.e., $\left|\chi_f\right| \ll 1$), then $\mu_f \approx \mu_0$ and hence Eq. (12) can be written as

$$F_m = \mu_0 V_p \frac{3\left(\chi_p - \chi_f\right)}{\left(\chi_p - \chi_f\right) + 3\left(\chi_f + 1\right)} (\mathbf{H_a}.\nabla)\mathbf{H_a}. \tag{13}$$

In similar fashion,

$$H_{in} = \frac{3}{\left(\chi_p - \chi_f\right) + 3} H_a, \tag{14}$$

and

$$m_{p,eff} = V_p \frac{3\left(\chi_p - \chi_f\right)}{\left(\chi_p - \chi_f\right) + 3} H_a. \tag{15}$$

In general, the effective dipole moment can be written as

$$m_{p,eff} = V_p f(H_a)\mathbf{H}_a, \tag{16}$$

where

$$f(H_a) = \begin{cases} \dfrac{3\left(\chi_p - \chi_f\right)}{\left(\chi_p - \chi_f\right) + 3}, & H_a < \left(\dfrac{\left(\chi_p - \chi_f\right) + 3}{3\chi_p}\right) M_{sp} \\ M_{sp}/H_a, & H_a < \left(\dfrac{\left(\chi_p - \chi_f\right) + 3}{3\chi_p}\right) M_{sp} \end{cases} \quad \left(|\chi_f| << 1\right), \tag{17}$$

and $H_a = |\mathbf{H}_a|$.

In this analysis, we assume that the magnet is infinitely extended in the y-direction, and therefore the $y-$ component of the magnetic field and force is considered as zero.

So, the intensity of the magnetic field and magnetic force has two components along $x-$ axis and $z-$ axis. The resultant of the intensity of the magnetic field is written as

$$\mathbf{H}_a = H_{ax}(x,z)\hat{x} + H_{az}(x,z)\hat{z} \tag{18}$$

and similarly, the resultant of the magnetic force is written as

$$\mathbf{F}_m = F_{mx}(x,z)\hat{x} + F_{mz}(x,z)\hat{z}, \tag{19}$$

where

$$F_{mx}(x,z) = \mu_0 V_p f(H_a)\left[H_{ax}(x,z)\frac{\partial H_{ax}(x,z)}{\partial x} + H_{az}(x,z)\frac{\partial H_{ax}(x,z)}{\partial z}\right], \tag{20}$$

and

$$F_{mx}(x,z) = \mu_0 V_p f(H_a)\left[H_{ax}(x,z)\frac{\partial H_{az}(x,z)}{\partial x} + H_{az}(x,z)\frac{\partial H_{az}(x,z)}{\partial z}\right]. \tag{21}$$

2.1. Magnetic field and force of a cylindrical magnet

Equations (20) and (21) represent the magnetic force for a rectangular cylinder whose y-axis extends infinitely and components of the magnetic force act along $x-$ and $z-$ axis. The calculation will be different for the case of cylindrical magnet. The cylindrical coordinate system (r', ϕ') with origin as the center of the magnet is shown in **Figure 4**.

Components of the intensity of the magnetic field in cylindrical coordinates are written as

$$H_{r'}\left(r', \phi'\right) = \frac{M_s R_{mag}^2}{2r'^2}\cos(\phi'), \tag{22}$$

and

$$H_{\phi'}\left(r',\phi'\right) = \frac{M_s R_{mag}^2}{2r'^2}\sin(\phi'). \tag{23}$$

Again, the geometry of the microvessel and flow nature is considered as the Cartesian coordinate system, and so it is necessary to convert the cylindrical coordinate system into the general Cartesian coordinate systems, say (x', z'); then, the intensity of the magnetic field in the Cartesian coordinate system is written as:

$$H_{x'}\left(x',z'\right) = \frac{M_s R_{mag}^2}{2}\frac{\left(x'^2 - z'^2\right)}{\left(x'^2 + z'^2\right)^2}, \tag{24}$$

$$H_{z'}\left(x',z'\right) = \frac{M_s R_{mag}^2}{2}\frac{2x'z'}{\left(x'^2 + z'^2\right)^2}, \tag{25}$$

and the corresponding field gradients is written as:

$$\frac{\partial}{\partial x'}\left(H_{x'}\left(x',z'\right)\right) = \frac{M_s R_{mag}^2}{2}\frac{x'\left(3z'^2 - x'^2\right)}{\left(x'^2 + z'^2\right)^3}, \tag{26}$$

$$\frac{\partial}{\partial z'}\left(H_{z'}\left(x',z'\right)\right) = \frac{M_s R_{mag}^2}{2}\frac{z'\left(z'^2 - 3x'^2\right)}{\left(x'^2 + z'^2\right)^3}, \tag{27}$$

$$\frac{\partial}{\partial z'}\left(H_{x'}\left(x',z'\right)\right) = \frac{M_s R_{mag}^2}{2}\frac{z'\left(z'^2 - 3x'^2\right)}{\left(x'^2 + z'^2\right)^3}, \tag{28}$$

$$\frac{\partial}{\partial z'}\left(H_{z'}\left(x',z'\right)\right) = M_s R_{mag}^2\frac{x'\left(x'^2 - 3z'^2\right)}{\left(x'^2 + z'^2\right)^3}. \tag{29}$$

Finally, components of the magnetic field from Eqs. (20) and (21) with the help of expression for intensity of the magnetic fields (Eqs. (26)–(29)) in a fixed Cartesian coordinate system (x, z) with respect to the microvessel are written as:

$$F_{mx}(x,z) = -\mu_0 V_p f(H_a) M_s^2 R_{mag}^4 \frac{(x+d)}{2\left[(x+d)^2 + z^2\right]^3}, \tag{30}$$

and

$$F_{mx}(x,z) = -\mu_0 V_p f(H_a) M_s^2 R_{mag}^4 \frac{z}{2\left[(x+d)^2 + z^2\right]^3}, \tag{31}$$

where d represents the distance between the axis of magnet and the carrier particle.

This magnetic force is further simplified for noninvasive magnetic drug targeting. In this procedure, the treatment takes place in such a way that there are no breaks in the skin, and there are no contacts with skin break or mucosa, or internal body cavity beyond a natural or artificial body orifice. In this case, the magnet is placed outside the body, and the distance from the magnet to blood vessel is much larger than the diameter of the blood vessel, that is, $d \gg x$. Furthermore, we used magnetite (Fe_3O_4) as a magnetic nanoparticle, which is highly biocompatible and for that the susceptibility of the magnetic particle is much larger, that is, $\chi_{mp} \gg 1$. Based on above two assumptions, the magnetic force components are simplified to:

$$F_{mx}(x,z) = -\frac{3\mu_0 V_p M_s^2 R_{mag}^4}{2\left(d^2+z^2\right)^3}, \tag{32}$$

and

$$F_{mx}(x,z) = -3\mu_0 V_p M_s^2 R_{mag}^4 \frac{z}{2\left(d^2+z^2\right)^3}, \tag{33}$$

3. Fluidic force on the particle

Fluidic force mainly represents the drug force acts on the particle and it depends on the velocity of the fluid. There are several phenomena, and studies have been carried out to understand the blood flow though artery. Blood is a marvelous fluid. Blood consists of a suspension of red blood cells (erythrocytes), white blood cells (leukocytes), and platelets in an aqueous solution (plasma). All contain different density. The plasma is a transparent, slightly yellowish fluid, and its density is about 1.035 gm/ml which can modeled as a Newtonian fluid. The red blood cells are dominant particulate matter in blood with about 40–45% by volume of the whole blood, and it plays a vital role in changing the viscosity of the blood. In general, we use the term "hematocrit" to specify the volume percentage of red cells and entrained plasma. Platelets are much smaller than red or white cells, and so it is not so significant for the nature of blood while it plays a vital role in the formation of blood clots which may severely interfere with the flow.

3.1. Fluidic force for single-phase flow (Newtonian fluid)

We start with a simple mathematical model for drag force. Here we considered blood as a Newtonian fluid. This nature of the blood flow appears mainly for vessel of radius more than 1500 μm. The flow nature is well discussed by Furlani and Furlani [23]. Using Stokes' approximation, the fluidic force for the Newtonian fluid is written as:

$$F_f = -6\pi\eta R_{cp}\left(v_{cp} - v_f\right), \tag{34}$$

where v_{cp} is the velocity of the carrier particle, v_f is the velocity of the fluid, and η is the viscosity of the fluid. First, we are interested to find the drag force for Newtonian fluid.

http://dx.doi.org/10.5772/intechopen.73678

Assume that the blood vessel is cylindrical and the flow is laminar, fully developed, and symmetric about the axis. So, the velocity profile is function of r (along the axis), and it is written as

$$v_f(r) = 2\overline{v}_f\left(1 - \left(\frac{r}{R_v}\right)^2\right), \tag{35}$$

where $\overline{v}_f$ is the average velocity of blood flow and R_v is the radius of the blood vessel.

With the help of Eq. (34), fluidic forces along the radial and axial radial direction are written as

$$F_{fr} = -6\pi\eta R_{cp} v_{p,x}, \tag{36}$$

and

$$F_{fz} = -6\pi\eta R_{cp}\left[v_{cp,z} - 2\overline{v}_f\left(1 - \left(\frac{r}{R_v}\right)^2\right)\right]. \tag{37}$$

3.2. Fluidic force for single-phase flow (non-Newtonian fluid model)

The radius of the microvessel plays a vital role on the nature of the blood flow. It is observed that the blood behaves as a non-Newtonian fluid due to low shear stress mainly when the diameter of the vessel is less than 1300 μm. The blood in microvessel is more complex due to the irregular geometry, mechanical behavior of blood, and its cellular constituents (mainly red blood cells). It is well established that the Casson models hold satisfactory for the blood flow through microvessel of diameter 130–1300 μm, while the Herschel-Bulkley model is more suitable for the microvessel of radius 20–100 μm (Mishra et al. [26]; Prier et al. [27]; Bugliarello and Sevilla [28]; Cokelet [29]; Merrill et al. [30]).

The drag force for the Newtonian fluid is different than the non-Newtonian fluid. The drag force for the spherical particle is written as

$$C_D = C_D(Q^*) = 24X(n)/Q^*, \tag{38}$$

where dynamic parameter $Q^* = \mathrm{Re}/(1 + Bi)^2$ with Reynolds number Re and Bingham number Bi and $X(n)$ is the rheological parameter.

For Casson fluid $X(n) = 1$, while Reynolds and Bringham number is written as (Shaw et al. [31])

$$\mathrm{Re}_{Casson} = \frac{v_{cp} d_{cp} \rho}{\nu}, \mathrm{Bi}_{Casson} = \sqrt{\frac{\rho\tau_y/\nu}{v_{cp}/d_{cp}}}, \tag{39}$$

where v_{cp} and d_{cp} are the reference velocity and diameter of the carrier particle ρ and ν are the density and kinematic viscosity of the fluid.

n	$X(n)$
0.1	1.354
0.2	1.413
0.3	1.458
0.4	1.442
0.5	1.420
0.6	1.382
0.7	1.320
0.8	1.240
0.9	1.140
1.0	1.002
1.2	0.827
1.4	0.569
1.6	0.390
1.8	0.261

Table 1. Value of rheological parameter $X(n)$ for different values of n.

For Herschel-Bulkley fluid, the rheological parameter $X(n)$ is defined in **Table 1**, and it is well correlated by the following analytical expression (Renaud et al. [32])

$$X(n) = 6^{(n-1)/2}\left\{\frac{3}{n^2+n+1}\right\}^{n+1}. \tag{40}$$

The Reynolds number and Birmingham number are written as

$$\mathrm{Re}_{HB} = \frac{v_{cp}^{2-n} d_{cp}^{n} \rho}{m}, \ \mathrm{Bi}_{HB} = \frac{\tau_y}{m\left(v_{cp}/d_{cp}\right)^{n}}, \tag{41}$$

where m is the Herschel-Bulkley model parameter.

3.3. Shape of the carrier particle

The shape of the carrier particles plays a significant role in magnetic drug targeting. A variety of nonspherical shapes including ellipsoids, discs, cylinders, hemispheres, cones, and red blood cell-like bioconcave discoids have been shown to have an impact on biological processes associated with the delivery of drugs (Geng et al. [33]; Canelas et al. [34]; Doshi and Mitragotri [35]; Enayati et al. [36]). Several experimental and laboratorial scientists have worked on the shape factor of particles in transport through the phagocytosis (Champion and Mitragotri [37]), circulation half-life (Geng et al. [33]), endocytosis (Gratton et al. [38]), targeting efficiency (Sutradhar et al. [39]), subsequent intracellular transport (Yoo et al. [40]), and vasculature (Decuzzi et al. [41]). Spherical and non-spherical carrier particles plays an imperative role on the passive drug targeting and bio-distribution of particles at the vascular level (Champion et al. [37]; Decuzzi et al. [42]; Mitragotri [43]; Fox et al. [44]). An excellent review of different shapes of carrier particles in drug delivery is given by Venkataraman et al. [45] where they discuss different experimental observations in this field and point out that understanding the implications of particle shapes would accelerate the development of next-generation drug delivery vehicles. It is noted that the shape of the magnetic nanoparticle does not change, and it is spherical in shape.

Due to the different shapes, the drag force on the carrier particle is modified as

$$C_D = C_D(Q^*) = 24k' X(n)/Q^*, \tag{42}$$

where k' is the shape constant which is defined as

$$k' = \begin{cases} 1, & \text{Sphere,} \\ \dfrac{\sqrt{\beta^2 - 1}}{\beta^{1/3}\ln\left(\beta + \sqrt{\beta^2 - 1}\right)}, & \text{Prolate ellipsoid,} \\ \dfrac{2\gamma}{3}\dfrac{1}{\ln\gamma + 0.193}, & \text{Cylinder,} \\ \dfrac{8\pi}{3}, & \text{Circular disk.} \end{cases} \tag{43}$$

The fluidic force on the carrier particle in a laminar flow is written as

$$F_f = -\frac{1}{2}\rho A_{cp} v_{cp}^2 C_D, \tag{44}$$

where A_{cp} is a reference cross-sectional area of the carrier particle.

With the help of Eq. (34), fluidic forces for the Casson fluid along the radial and axial radial direction are written as

$$F_{fr} = -12A_{cp}\eta k' \left[\left(\frac{\tau_y}{\eta}\right)^{1/2} + \left(\frac{v_{cp,x}}{d_{cp}}\right)^{1/2}\right]^2, \tag{45}$$

and

$$F_{fz} = -12A_{cp}\eta k' \left[\left(\frac{\tau_y}{\eta}\right)^{1/2} + \left(\frac{v_{cp,z} - \overline{v}_f}{d_{cp}}\right)^{1/2}\right]^2. \tag{46}$$

However, fluidic forces for the Herschel-Bulkley fluid along the radial and axial radial direction are written as

$$F_{fr} = -12A_{cp}X(n)k' \left[\tau_y + \eta\left(\frac{v_{cp,x}}{d_{cp}}\right)^n\right], \tag{47}$$

and

$$F_{fz} = -12A_{cp}X(n)k' \left[\tau_y + \eta\left(\frac{v_{cp,z} - \overline{v}_f}{d_{cp}}\right)^n\right]. \tag{48}$$

3.4. Fluidic force for two-phase flow

Due to the microscopic properties of the blood and interaction among its different particles (Blood cells, platelets) in plasma, the nature of blood flow leads a two-phase flow model with a core of rouleaux surrounded by a cell-depicted peripheral layer. Due to the translation, deformation, and rotation, RBCs accumulated near the axis of the vessel and followed a constant velocity while the plasma layer appears near the vessel wall and velocity profile follows a parabolic profile. A mathematical model for a two-phase fluid model has been discussed by Seshadri and Jaffrin [46] in which they have considered the outer layer as cell depleted, having a lower hematocrit than the core region. Later, several researchers are worked on this direction (Gupta et al. [47]; Srivastava [48]; Sankar and Lee [49]).

3.5. Glycocalyx layer and permeability of the microvessel

Permeability of the vessel is another important characteristic, which mainly influences the flow nature of the blood. Due to the wall permeability, the fluid flows laterally and across the vessel fenestrations/pores. Fluid flow through the porous medium is defined by the Darcy law and extended Darcy law or Brinkman Law. Microvessel walls consist mainly of endothelial cell. The vascular endothelium layer helps to regulate the material exchange between circulating blood and the body tissues. The mechanism of the endothelium cells modulates microvessel permeability. It is well known that the luminal surface of the vascular endothelium is lined with a glycocalyx, a layer of membrane-bound macromolecules and adsorbed plasma proteins. This glycocalyx layer is capable of reducing or restricting the plasma flow at the peripheral layer, near to the vessel wall. Moreover, the presence of the glycocalyx layer decreases the effective cross-sectional area of the vessel available for plasma and red cell motion, so that it would possibly cause an increase in flow resistance. The flow resistance highly depends on the thickness of the glycocalyx layer. The flow resistance of the glycocalyx layer is more higher for the thick layer with respect to the thin layer (Pries et al. [27]; Weinbaum et al. [50]; Sugihara-Seki and Fu [51]). It is observed that the permeability of the microvessel significantly changes from organ to organ with the location of the microvessel and is written in ascending order as brain < skin < skeletal muscle < lung < heart <gastrointestinal tract< glomerulus of kidney.

3.6. Porosity of the carrier particle

In recent years, researchers are focused on the development of porous materials as controlled drug delivery matrices due to its unique features such as stable uniform porous structure, high surface area, tunable pore sizes with narrow distribution, and well-defined surface properties (Sher et al. [52]; Shivanand and Sprockel [53]; Ahuja and Pathak [54]) (**Figure 5**). Due to these wide physical flexibilities, porous carriers have been used in pharmaceuticals for many purposes including development of novel drug delivery systems such as floating drug delivery system, sustained drug delivery system, and improvement of solubility of poorly soluble drugs (Sharma et al. [55]; Streubel et al. [56]). These materials possess larger amounts of nanopores that allow the inclusion of drugs (Wang et al. [57]). Also, these features allow them to adsorb drugs and release them in a more reproducible and predictable manner. The use of mesoporous,

Figure 5. Electron microscopy images of porous Cu_2O nanospheres (Zhu [60]).

microporous, and nanoporous carriers for drug delivery is a part of growing research (Song et al. [58]; Andersson [59]).

4. Particle-particle interaction

In a small vessel where diameter of the vessel is less than 500 μm, it is observed that the apparent viscosity of blood is significantly decreased. This reduction of the apparent blood viscosity with decreasing diameter is continued down to the diameter of approximately 10 μm. This phenomenon was first observed by Martini et al. [61] and at the same time by Fahraeus and Lindqvist [62] during their vitro experiment to measure apparent viscosity of blood in a small narrow glass tube. This phenomenon is called as the Fahraeus-Lindqvist effect after their name. The reduction of the apparent viscosity of blood mainly appeared due to the rotation and displacement of the deformable cells (mainly red blood cells) toward the tube axis so that a cell-depleted region (mainly plasma) is formed near the wall. The effect of the cell-depleted region does not significantly influence the apparent viscosity of blood for large vessels ($d > 500$ μm) as the width of the cell-depleted layer is much smaller than the diameter of the vessel. Pries et al. [63] compiled literature data on relative blood viscosity in tube flow in vitro and obtained the empirical relationship between the relative apparent viscosity and tube diameter for red cell suspensions with a hematocrit of 45% in glass tubes.

$$\left(\frac{\mu_{app}}{\mu}\right)_{0.45} = 220e^{-1.3d} + 3.2 - 2.44e^{-0.06d^{0.645}}, \tag{49}$$

where μ_{app} is the apparent viscosity, μ is the viscosity of the plasma, and d (in μm) is the diameter of the tube. The apparent viscosity of blood for the medium-to-high flow velocities (above 50 tube diameters/s) is written as

$$\frac{\mu_{app}}{\mu} = 1 + \left[\left(\frac{\mu_{app}}{\mu}\right)_{0.45} - 1\right]\frac{(1 - H_d)^C - 1}{(1 - 0.45)^C - 1}, \tag{50}$$

where

$$C = \left(0.8 + e^{-0.075d}\right)\left(-1 + \frac{1}{1 + 10(d/10)^{12}}\right) + \frac{1}{1 + 10(d/10)^{12}}. \tag{51}$$

Typical flow patterns of red blood cells in narrow tubes with different diameters are shown in **Figure 6**. In all cases, at the wall where the highest shear stress occurs (wall shear stress), the local viscosity of blood is the same as the viscosity of plasma ($\sim 1.2\ cP$), which reduces the wall shear stress and flow resistance.

Red blood cells are gathered near the axis of the vessel and so a cell-free region or low hematocrit region appears near the wall of the vessel. Red blood cells flow faster than the cell-depleted wall layer which mainly contains plasma and due to that the average velocity at the core region (RBCs' region) is much higher than the outer region (plasma region). This phenomenon leads to a reduction of the tube hematocrit as compared to the discharged hematocrit. A simple schematic diagram of this model is given in **Figure 6** where a steady laminar flow of suspension of cells through a tube to or from a reservoir is considered. The problem leads to conservation of mass, which is written as

$$H_T/H_d = U_S/U_C, \tag{52}$$

where H_T and H_d represent the volume concentration of cells in the tube and in reservoir, respectively. U_S and U_C are the mean velocities of the suspension and cells, respectively. In narrow tubes, $U_S < U_C$ which gives $H_T < H_d$ and it implies that H_T/H_d decreases with decrease in tube diameter. This phenomenon is called Fahraeus effect. A further decrease in tube diameter below approximate 30 μm and less, we find an increase in H_T/H_d, which known as inverse Fahraeus effect (**Figure 7**). The dependence of H_T/H_d on tube diameter D (in μm) and discharge hematocrit H_d is compiled by Pries et al. [64] and is written as

$$\frac{H_T}{H_d} = H_d + (1 - H_d)\left(1 + 1.7e^{-0.415d} - 0.6e^{-0.011d}\right). \tag{53}$$

Consider a Newtonian fluid flow through the vessel as cylindrical tube of diameter d. Consider the viscosity at the core region and outside the core region (or at plasma region) as μ and μ_0, respectively. Considered the continuity of velocity and stresses at the interface of the core and

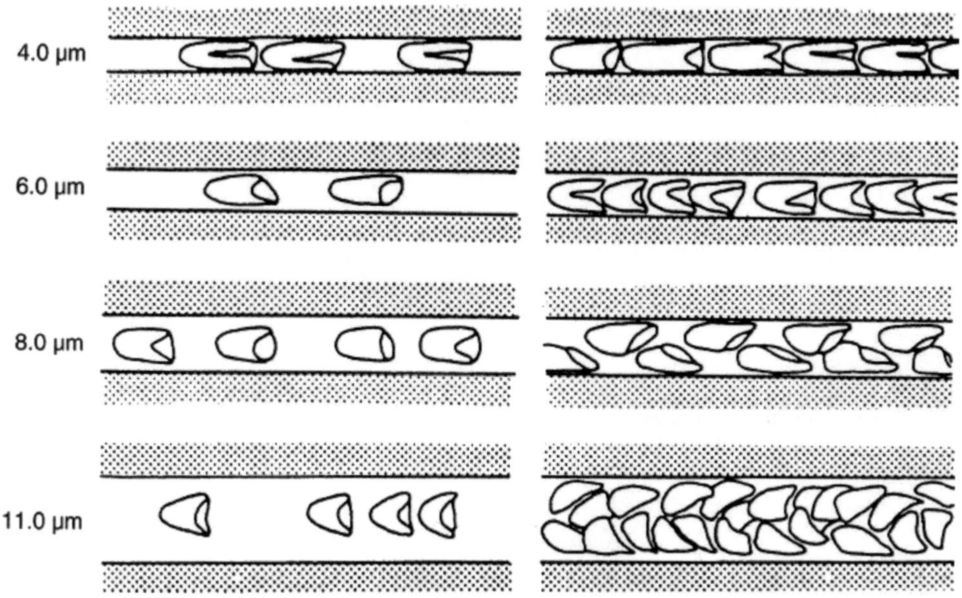

Figure 6. Schematic drawings of red cells flowing through narrow glass tubes. The flow is from right to left. The left panel is at low hematocrit, and the right panel is at high hematocrit. Single-file flow is always present in smaller tubes, whereas transition to "zipper" or multi-file flow may occur at higher hematocrit in tubes larger than approximately 7 μm (Gaehtgens [66]).

outer region while a non-slip condition on the tube of the wall. Using the abovementioned assumption, the velocity profile of the fluid as a function of radial coordinates r is written as

$$u = \begin{cases} \frac{P_z}{4\mu}\left[\left(\frac{d}{2}\right)^2 - r^2\right], & \lambda d^2 \le r \le d/2, \\ \frac{P_z}{4\mu_0}\left[\left(\frac{\lambda d}{2}\right)^2 - r^2\right] + \frac{P_z}{4\mu}\left(\frac{d}{2}\right)^2(1-\lambda^2), & 0 \le r \le d/2, \end{cases} \tag{54}$$

where P_z represents the pressure gradient along the tube axis, λ is a constant which lies between 0 and 1, and λd is the diameter of the core region with d as the diameter of the cylinder (vessel).

Using Eqs. (52) and (53), the Fahraeus-Lindqvist effect and Fahraeus effects (Pries et al. [65]) are evaluated as

$$\mu_{app} = \frac{\mu}{1-\left(1-\frac{\mu}{\mu_0}\right)P_z^4}, \quad \frac{H_T}{H_d} = \frac{1}{2}\frac{1-P_z^4\left(1-\frac{\mu}{\mu_0}\right)}{1-P_z^2\left(1-\frac{\mu}{2\mu_0}\right)}. \tag{55}$$

In the limit of λ approaching 0, we have $H_T/H_d \to 1/2$ and $\mu_{app}/\mu \to 1$, and in the other way for the limit of λ approaching 1, we have $H_T/H_d \to 1$ and $\mu_{app}/\mu \to \mu_0/\mu$.

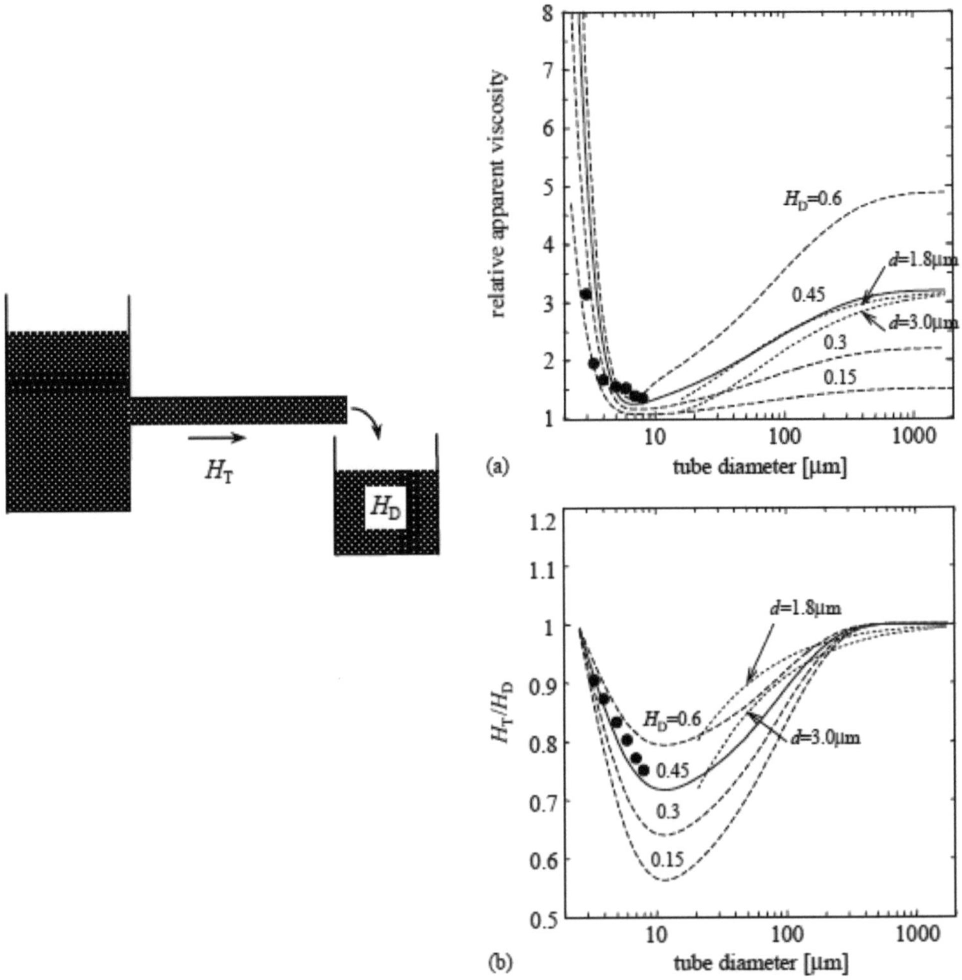

Figure 7. (a) Relative apparent viscosity of blood and (b) hematocrit ratio *HT/HD*, as a function of tube diameter. Solid and dashed curves, empirical fit to in vitro experimental data for blood flow in glass tubes (Eqs. (52), (53), and (55)); Dotted curves, prediction by the stacked-coins model Eq. (55); filled circles, theoretical predictions based on axisymmetric geometries at cell velocity = 1 mm/s (Secomb et al. [68]; Secomb [69]). Eqs. (52) and (55) at *HD* = 0.45 are represented by solid curves, corresponding to normal states.

4.1. Saffman lift force

Saffman lift force is mainly effective for small particles in the shear field. During drug targeting, we always used carrier particles with nanosizes, and so the particles experience a lift force perpendicular to the direction of the flow, which is known as Saffman lift force. The shear lift originates from the inertia effect in the viscous flow around the particle, and it can be defined as (Zheng and Silber-Li [67])

$$F_{saff} = 8k^{'}\mu\sqrt{\dot{\gamma}/\nu}R_{cp}^{2}(v_f - v_{cp}), \tag{56}$$

where $\dot{\gamma}$ is the shear rate of the fluid and k' is the shape factor.

5. Buoyant force

For the large artery of diameter more than 1000 μm, the buoyant force plays a vital role with the convective term due to the mass of the carrier particle. The buoyant force acting on the carrier particle during its motion in the microvessel is written as

$$F_b = -V_{cp}\left(\rho_{cp} - \rho_f\right)g\mathbf{x}, \tag{57}$$

where g is the gravitational acceleration and V_{cp} is the volume of the carrier particle.

6. Geometry of the microvessel

The geometry of the microvessel is not uniform throughout the cardiovascular system. Different geometries of the vessel have been observed in the cardiovascular system, such as bifurcated vessel, with atherosclerosis inside the vessel which is attached with the wall of the vessel, curved vessel, and so on. The flow phenomenon is very complex and interesting when blood flows through curved or bifurcated or stenosed artery. Sometimes secondary flow appears near the apex of the bifurcation vessel or near stenosis. In general, numerical simulations are performed to obtain better insights into the theoretical analysis with computational fluid dynamics (Wang et al. [57]; Pourmehran et al. [17]).

7. Inertia force

Inertia force is playing an important role in blood flow through a vessel with a large diameter. In general we may use carrier particles with large radius, which can easily capture the tumor position. In general multiplication of mass of the carrier particle (which is actually same as the velocity of fluid) and acceleration of carrier particles (which is same as the acceleration of the blood flow in vessel) is not negligible, and it gives a significant influence in the magnetic drug targeting.

8. Equation of motion

Equation of the motion is followed by Newton's second law of motion. According to this law, the total force on any particle is equal to the multiplication of mass and acceleration of the particle. According to this law, the equation of motion is written as

$$m_{cp}\frac{dv_{cp}}{dt} = Sum of\ All forces = F_m + F_f + F_{saff} + F_b, \tag{58}$$

where m_{cp} is the mass of the carrier particle.

Equation (58) represents the force equation in the presence of inertia force. While in the absence of inertia force, when the diameter of the vessel is very small, then the right part of the Eq. (58) is negligible, and the equation of motion can be written as

$$0 = Sum\,of\ All\,forces = F_m + F_f + F_{saff} + F_b. \tag{59}$$

9. Conclusion and future work

Magnetic drug targeting is one of the very useful and biocompatible noninvasive methods. Still there are many scopes left to study and build different mathematical models. There are several factors that directly and indirectly act on the drug targeting and influence the trajectory of the carrier particle. Moreover, here the main focus is on the magnetic drug targeting in the cardiovascular, more particularly in the microvessel. Several experimental works have been done in this direction, which can be implemented as mathematical models. Also, magnetic drug targeting is effective for other human systems such as renal system, gastrointestinal tract, and so on.

Author details

Sachin Shaw

Address all correspondence to: shaws@biust.ac.bw

Department of Mathematics and Statistical Sciences, Botswana International University of Science and Technology, Palapye, Botswana

References

[1] Furlani EP. Magnetic biotransport: Analysis and applications. Materials. 2010;**3**:2412-2446

[2] Moroz P, Jones SK, Gray BN. Magnetically mediated hyperthermia: Current status and future directions. International Journal of Hyperthermia. 2002;**18**:267-284

[3] Hergt R, Duzt S. Magnetic parti therapycle hyperthermia-biophysical limitations of a visionary tumor. Journal of Magnetism and Magnetic Materials. 2007;**311**:187-192

[4] Gupta AK, Gupta M. Synthesis and surface engineering of iron oxide nanoparticles for biomedical applications. Biomaterials. 2005;**26**:3995-4021

[5] Shaw S, Murthy P. Magnetic drug targeting in the P ermeable blood vessel—The effect of blood rheology. Journal of Nanotechnology in Engineering and Medicine. 2010;**1**: 021001

[6] Shaw S, Murthy P, Pradhan SC. Effect of non-Newtonian characteristics of blood on magnetic targeting in the impermeable microvessel. Journal of Magnetism and Magnetic Materials. 2010;**322**:1037-1043

[7] Shaw S, Murthy P. The effect of shape factor on the magnetic targeting in the permeable microvessel with two-phase casson fluid model. Journal of Nanotechnology in Engineering and Medicine. 2011;**2**:041003 (8 Pages)

[8] Kircher MF, Mahmood U, King RS, Weissleder R, Josephon L. A multimodal nanoparticle for preoperative magnetic resonance imaging and intraoperative optical brain nanoparticle for preoperative magnetic resonance imaging and intraoperative optical brain. Cancer Research. 2003;**63**:8122-8125

[9] Sahoo Y, Goodarzi A, Swihart MT, Ohulchanskyy TY, Kaur N, Furlani EP, Prasad PN. Ferrofluid of magnetite nanoparticles: fluorescence labeling and magnetophoretic control. Journal of Physical Chemistry B. 2005;**109**:3879-3885

[10] Sosnovik DE, Nahrendorf M, Weissleder R. Magnetic nanoparticles for MR imaging: Agents, techniques and cardiovascular applications. Basic Research in Cardiology. 2008; **103**:122-130

[11] Prasad PN, Introduction to Biophotonics. NJ, USA: John Wiley & Sons: Hoboken; 2005

[12] Levy L, Sahoo Y, Kim KS, Bergey EJ, Prasad PN. Nanochemistry: Synthesis and characterization of multifunctional nanoclinics for biological applications. Chemistry of Materials. 2002;**14**:3715-3721

[13] Medarova Z, Pham W, Kim Y, Dai G, Moore A. In vivo imaging of tumor response to therapy using a dual-modality imaging strategy. International Journal of Cancer. 2006; **118**:2796-2802

[14] Lubbe AS, Alexiou C Bergenmann C. Clinical application of magnetic drug targeting, The Journal of Surgical Research;**95**(2):200-206

[15] Alexiou C, Tietze R, Schreiber E, Jurgons R, Richter H, Rahn H, Odenbach S, Lyer S. Cancer therapy with drug loaded magnetic nanoparticles—Magnetic drug targeting. Journal of Magnetism and Magnetic Materials. 2011;**323**:1404-1407

[16] Jurgons R, Seliger C, Hilpert A, Trahms L, Odenbach S, Alexiou C. Drug loaded magnetic nanoparticles for cancer therapy. Journal of Physics: Condensed Matter. 2006;**18**:S2893-S2902

[17] Pourmehran O, Gorji TB, Gorji-Bandpy M. Magnetic drug targeting through a realistic model of human tracheobronchial airways using computational fluid and particle dynamics. Biomechanics and Modeling in Mechanobiology. 2016;**15**:1355-1374

[18] Fabrizio M, Francois L. Active targeting with particulate drug carriers in tumor therapy: Fundamentals and recent progress. Drug Discovery Today. 2004;**9**:219-228

[19] Berryl CC, Curtis AS. Functionalisation of magnetic nanoparticles for appications in biomedicine. Journal of Physics D: Applied Physics. 2003;**36**:R198-R206

[20] Fan J-B, Huang C, Jiang L, Wang S. Nanoporous microspheres: From controllable synthesis to healthcare applications. Journal of Materials Chemistry B. 2013;**1**:2222-2235

[21] Arrueboa M, Fernandez-Pachecoa R, Ibarraa RM, Santamaria J. Magnetic nanoparticles for drug delivery, Nanotoday. 2017;**2**:22-32

[22] Lunnoo T, Puangmali T. Capture efficiency of biocompatible magnetic nanoparticles in arterial flow: A computer simulation for magnetic drug targeting, nanoscale research letters. 2015;**10**:426 (10 pages)

[23] Furlani EJ, Furlani EP. A model for predicting magnetic targeting of mulfunctional particles in the microvasculature. Journal of Magnetism and Magnetic Materials. 2007; **312**:187-193

[24] Jones TP. Electomagnetics of Particles. Cambridge, UK: Cambridge University Press; 1995

[25] Furlani EP. Permanent Magnet and Electomechanical Devices : Materials. Analysis and Applications, New York: Academic Press; 2001

[26] Mishra JC, Patra MK, Sahu BK. Unsteady flow of blood through narrow blood vessels—A mathematical analysis. Computers in Mathematical Application. 1992;**24**:19-31

[27] Pries AR, Secomb TW, Gaehtgens P. The endothelial surface layer. Pflügers Archiv: European Journal of Physiology. 2000;**440**:653-666

[28] Bugliarello G, Sevilla J. Velocity distribution and other characteristics of steady and pulsatile blood flow in fine glass, Biorheology. 1970;**7**:85-107

[29] Cokelet GR. The rheology of human blood: In Biomechanics, YC Fung, editor. New Jersey: Prentice Hall, Englewood Cliffs; 1972

[30] Merrill FW, Benis AM, Gilliland ER, Sherwood TK, Salzman EW. Pressure-flow relations of human blood in hollow fibers at low flow rates. Journal of Applied Physiology. 1965; **20**:954-967

[31] Shaw S, Sutradhar A, Murthy P. Permeability and stress-jump effects on magnetic drug targeting in a permeable microvessel using Darcy model. Journal of Magnetism and Magnetic Materials. 2017;**429**:227-235

[32] Renaud M, Mauret E, Chhabra RP. Power law fluid flow over a sphere: Average shear rate and drag coefficient. The Canadian Journal of Chemical Engineering. 2004;**82**:1066-1070

[33] Geng Y, Dalhaimer P, Cai S, Tsai R, Tewari M, Minko T, Discher DE. Shape effect of filaments versus spherical particles in flow and drug delivery. Nature Nanotechnology. 2007;**2**:249-255

[34] Canelas D, Herlihy KP, DeSimone JM. Top-down particle fabrication : Control of size and shape for diafnostic imaging and drug delivery. Wiley Interdisciplinary Reviews. Nanomedicine and Nanobiotechnology. 2009;**1**:391-404

[35] Doshi N, Mitragotri S. Designer biomaterials for nanomedicine. Advanced Functional Materials. 2009;**19**:3843-3854

[36] Enayati M, Ahmed Z, Stride E, Edirisinghe M. Preparation of polymeric carriers for drug delivery with different shape and size using an electric jet. Current Pharmaceutical Biotechnology. 2009;**10**:600-608

[37] Champion JA, Katare YK, Mitragotri S. Particle shape: A new design parameters for micro- and nanoscale drug delivery carriers. Journal of Controlled Release. 2007;**121**:3-9

[38] Gratton SE, Ropp PA, Pohlhaus PD, Luft JC, Madden VJ, Napier ME, Desimore JM. The effect of particle design on cellular internalization pathways. Proceedings of the National Academy of Sciences. 2008;**105**:613-618

[39] Sutradhar A, Murthy P, Shaw S. Influence of the inertia on magnetic drug targeting in microvessel-Casson model. Journal of Nanofluids. 2016;**5**:1-7

[40] Yoo J, Doshi N, Mitragotri S. Endocytosis and intracellular distribution of PGlA particles in endothelial cells: Effect of particle geometry. Macromolecular Rapid Communications. 2009;**31**:142-148

[41] Decuzzi P, Godin B, Tanaka T, Lee SY, Chiappini C, Liu X, Ferrari M. Size and shape effects in the bio-distribution of intravascularly injected particles. Journal of Controlled Release. 2010;**141**:320-327

[42] Deccuzi P, Pasqualini R, Arap W, Ferrari M. Intravascular delivery of particulate systems: Does geometry really matter? Pharmaceutical Research. 2009;**26**:235-243

[43] Mitragotri S. In drug delivery, shape does matter. Pharmaceutical Research. 2009;**26**:232-234

[44] Fox ME, Szoka FC, Frechet JM. Soluble polymer carriers for the treatment of cancer: The importance of molecular architecture. Accounts of Chemical Research. 2009;**42**:1141-1151

[45] Venkataraman S, Hedrick JL, Ong ZY, Yang C, Rachel PI, Hammond PT, Yang YY. The effects of polymeric nanostructure shape on drug delivery. Advanced Drug Delivery Reviews. 2011;**63**:1228-1246

[46] Seshadri V, Jaffrin MY. Anomalous effects in blood flow through narrow tubes: A model. INSERM-Eutomech. 1977;**92**:265-282

[47] Gupta BB, Nigam KM, Jaffrin MY. A three-layer semi-empirical model for flow of blood and other particulate suspensions through narrow tubes. Journal of Biomechanical Engineering. 1982;**104**:129-135

[48] Srivastava VP. Two-phase model of blood flow through stenosed tubes in the presence of a peripheral layer: Applications. Journal of Biomechanics. 1996;**29**:1377-1382

[49] Sankar DS, Lee U. Two-phase non-linear model for the blood flow through stenosed blood vessels. Journal of Mechanical Science and Technology. 2007;**21**:678-689

[50] Weinbaum S, Zhang X, Han Y, Vink H, Cowin SC. Mechanotransduction and flowacross the endothelial glycocalyx. PNAS. 2003;**100**:7988-7995

[51] Sugihara-Seki M, Fu BM. Blood flow and permeability in microvessels. Fluid Dynamics Research. 2005;**37**:82-132

[52] Sher P, Insavle G, Porathnam S, Pawar AP. Low density porous carrier drug adsorption and release study by response surface methodology using different solvents. International Journal of Pharmaceutics. 2007;**331**:72-83

[53] Shivananda P, Sprockel OL. A controlled porosity drug delivery system. International Journal of Pharmaceutics. 1998;**167**:83-96

[54] Ahuja G, Pathak K. Porous carriers for controlled/modulated drug delivery. Indian Journal of Pharmaceutical Sciences. 2009;**71**:500-507

[55] Sharma S, Sher P, Badve S, Atmaram PP. Adsorption of meloxicam on porous calcium silicate: Characterisation and tablet formulation. AAPS PharmSciTech. 2005;**6**:E618-E625

[56] Streubel A, Sipemann J, Bodmeier R. Floating matrix tablets based on low density foam powder: Effects of formulation and processing parameters on drug release. European Journal of Pharmaceutical Sciences. 2003;**18**:37-45

[57] Wang S, Liu H, Xu W. Hydrodynamic modelling and CFD simulation of ferrofluids flow in magnetic targeting drug delivery. International Journal of Computational Fluid Dynamics. 2008;(10):659-667

[58] Song SW, Hidajat K, Kawi S. Functionalized SAB-15 material as carrier from controlled drug delivery: Infl uence of surface properties on matrix drug interactions. Langmuir. 2005;**21**:9568-9575

[59] Andersson J, Rsenhoim J, Areva S, Linden M. Influences of material characteristics on ibuprofen drug loading and release profiles from ordered micro- and mesoporous silica matrices. Chemistry of Materials. 2004;**16**:4160-4167

[60] Zhu JJ. Nanotechweb, 19 Jan 2009 [Online] Available: http://nanotechweb.org/cws/article/lab/37301

[61] Martini P, Pierach A, Schreyer E. Die Stromung des Blutes in eigen Gefassen Eine Abweichung vom Poiseuille'schne Gesetz. Dtsch Arch Klin Med. 1930;**169**:212-222

[62] Fahraeus R, Lindqvist T. The viscosity of the blood in narrow capillary tubes. The American Journal of Physiology. 1931;**96**:562-568

[63] Pries AR, Neuhaus D, Gaehtgens P. Blood viscosity in tube flow: Dependence on diameter and hematocrit. American Journal of Physiology. 1992;**263**:H1770-H1778

[64] Prier AR, Secomb TW, Gaehtgens P. Biophysical aspects of blood flow in the microvessel. Cardiovascular Research. 1996;**32**:654-667

[65] Pries AR, Secomb TW, Gaehtgens P, Gross JF. Blood flow in microvascular networks: Experiments and simulation. Circulation Research. 1990;**67**:826-834

[66] Gaehtgens P. Flow of blood through narrow capillaries: Rheological mechanisms determining capillary hematocrit and apparent viscosity. Biorheology. 1980;**17**:183-189

[67] Zheng X, Silber-Li Z. The influence of Saffman lift force on nanoparticle concentration distribution near a wall. Applied Physics Letters. 2009;**95**:124105 (3 Pages)

[68] Secomb TW, Shalak R, Ozkaya N, Gross JF. Flow of axisymmetric red blood cells in narrow capillaries. Journal of Fluid Mechanics. 1986;**163**:405-423

[69] Secomb TW. Mechanics of Blood Flow in the Microcirculation, in Biological Fluid Dynamics. London: Company of Biologists; 1995:305-321